# Guidelines for drinking-water quality

SECOND EDITION

*Volume 1*
*Recommendations*

World Health Organization
Geneva
1993

WHO Library Cataloguing in Publication Data

Guidelines for drinking-water quality. — 2nd ed.
    Contents: v. 1. Recommendations
    1.Drinking water — standards
    ISBN 92 4 154460 0 (v. 1) (NLM Classification: WA 675)

The World Health Organization welcomes requests for permission to reproduce or translate its publications, in part or in full. Applications and enquiries should be addressed to the Office of Publications, World Health Organization, Geneva, Switzerland, which will be glad to provide the latest information on any changes made to the text, plans for new editions, and reprints and translations already available.

© **World Health Organization 1993**

Publications of the World Health Organization enjoy copyright protection in accordance with the provisions of Protocol 2 of the Universal Copyright Convention. All rights reserved.

The designations employed and the presentation of the material in this publication do not imply the expression of any opinion whatsoever on the part of the Secretariat of the World Health Organization concerning the legal status of any country, territory, city or area or of its authorities, or concerning the delimitation of its frontiers or boundaries.

The mention of specific companies or of certain manufacturers' products does not imply that they are endorsed or recommended by the World Health Organization in preference to others of a similar nature that are not mentioned. Errors and omissions excepted, the names of proprietary products are distinguished by initial capital letters.

PRINTED IN FRANCE

93/9625 — Sadag — 8000

# Contents

Preface ....... vii

Acknowledgements ....... ix

Acronyms and abbreviations used in the text ....... x

## 1. Introduction ....... 1

    1.1 General considerations ....... 2
    1.2 The nature of the guideline values ....... 4
    1.3 Criteria for the selection of health-related
        drinking-water contaminants ....... 6

## 2. Microbiological aspects ....... 8

    2.1 Agents of significance ....... 8
        2.1.1 Waterborne infections ....... 8
        2.1.2 Orally transmitted infections of high priority ....... 8
        2.1.3 Opportunistic and other water-associated pathogens ....... 9
        2.1.4 Toxins from *Cyanobacteria* ....... 9
        2.1.5 Nuisance organisms ....... 11
        2.1.6 Persistence in water ....... 12
        2.1.7 Infective dose ....... 13
        2.1.8 Guideline values ....... 13

    2.2 Microbial indicators of water quality ....... 14
        2.2.1 Introduction ....... 14
        2.2.2 General principles ....... 15
        2.2.3 *Escherichia coli* and the coliform bacteria ....... 15
        2.2.4 Faecal streptococci ....... 17
        2.2.5 Sulfite-reducing clostridia ....... 18
        2.2.6 Coliphages and other alternative indicators ....... 18
        2.2.7 Methods of detection ....... 18

|   |   |
|---|---|
| 2.3 Recommendations | 20 |
|    2.3.1 General principles | 20 |
|    2.3.2 Selection of treatment processes | 20 |
|    2.3.3 Treatment objectives | 21 |
|    2.3.4 Guideline values | 21 |
| 2.4 Monitoring | 24 |
|    2.4.1 Approaches and strategies | 24 |
|    2.4.2 Sampling frequencies | 25 |
|    2.4.3 Sampling procedures | 26 |
|    2.2.4 Surveillance programme requirements | 28 |

## 3. Chemical aspects    30

|   |   |
|---|---|
| 3.1 Background information used | 30 |
| 3.2 Drinking-water consumption and body weight | 30 |
| 3.3 Inhalation and dermal absorption | 31 |
| 3.4 Health risk assessment | 31 |
|    3.4.1 Derivation of guideline values using a tolerable daily intake approach | 32 |
|    3.4.2 Derivation of guideline values for potential carcinogens | 35 |
| 3.5 Mixtures | 39 |
| 3.6 Summary statements | 39 |
|    3.6.1 Inorganic constituents | 39 |
|    3.6.2 Organic constituents | 57 |
|    3.6.3 Pesticides | 75 |
|    3.6.4 Disinfectants and disinfectant by-products | 93 |
| 3.7 Monitoring | 105 |
|    3.7.1 Design of a sampling programme | 106 |
|    3.7.2 Sample collection | 110 |
|    3.7.3 Analysis | 111 |

## 4. Radiological aspects    114

|   |   |
|---|---|
| 4.1 Introduction | 114 |
|    4.1.1 Environmental radiation exposure | 114 |
|    4.1.2 Potential health consequences of radiation exposure | 115 |
|    4.1.3 Recommendations | 115 |
| 4.2 Application of the reference level of dose | 116 |
|    4.2.1 Analytical methods | 118 |
|    4.2.2 Strategy for assessing drinking-water | 119 |
|    4.2.3 Radon | 120 |

## 5. Acceptability aspects — 122

   5.1 Introduction — 122
   5.2 Summary statements — 123
      5.2.1 Physical parameters — 123
      5.2.2 Inorganic constituents — 124
      5.2.3 Organic constituents — 128
      5.2.4 Disinfectants and disinfectant by-products — 129

## 6. Protection and improvement of water quality — 131

   6.1 General considerations — 131
   6.2 Selection and protection of water sources — 132
   6.3 Treatment processes — 132
      6.3.1 Pre-treatment — 133
      6.3.2 Coagulation, flocculation, and sedimentation — 134
      6.3.3 Rapid and slow sand filtration — 134
      6.3.4 Disinfection — 135
      6.3.5 Fluoride removal — 136
   6.4 Choice of treatment — 136
   6.5 Distribution networks — 137
   6.6 Corrosion control — 138
      6.6.1 Introduction — 138
      6.6.2 Basic considerations — 138
      6.6.3 Effect of water composition — 139
      6.6.4 Corrosion of pipe materials — 140
      6.6.5 Microbiological aspects of corrosion — 141
      6.6.6 Corrosion indices — 141
      6.6.7 Strategies for corrosion control — 142
   6.7 Emergency measures — 142

**Bibliography** — 144

**Annex 1. List of participants in preparatory meetings** — 149
**Annex 2. Tables of guideline values** — 172

**Index** — 183

# Preface

In 1984 and 1985, the World Health Organization (WHO) published the first edition of *Guidelines for drinking-water quality* in three volumes. The development of these guidelines was organized and carried out jointly by WHO headquarters and the WHO Regional Office for Europe (EURO).

In 1988, the decision was made within WHO to initiate the revision of the guidelines. The work was again shared between WHO headquarters and EURO. Within headquarters, both the unit for the Prevention of Environmental Pollution (PEP) and the ILO/UNEP/WHO International Programme on Chemical Safety (IPCS) were involved, IPCS providing a major input to the health risk assessments of chemicals in drinking-water.

The revised guidelines are being published in three volumes. Guideline values for various constituents of drinking-water are given in Volume 1, *Recommendations* together with essential information required to understand the basis for the values. Volume 2, *Health criteria and other supporting information,* contains the criteria monographs prepared for each substance or contaminant; the guideline values are based on these. Volume 3, *Surveillance and control of community supplies,* is intended to serve a very different purpose; it contains recommendations and information concerning what needs to be done in small communities, particularly in developing countries, to safeguard their water supplies.

The preparation of the current edition of the *Guidelines for drinking-water quality* covered a period of four years and involved the participation of numerous institutions, over 200 experts from nearly 40 different developing and developed countries and 18 meetings of the various coordination and review groups. The work of these institutions and scientists, whose names appear in Annex 1, was central to the completion of the guidelines and is much appreciated.

For each contaminant or substance considered, a lead country prepared a draft document evaluating the risks for human health from exposure to the contaminant in drinking-water. The following countries prepared such evaluation documents: Canada, Denmark, Finland, Germany, Italy, Japan, Netherlands, Norway, Poland, Sweden, United Kingdom of Great Britain and Northern Ireland and United States of America.

Under the responsibility of a coordinator for each major aspect of the guidelines, these draft evaluation documents were reviewed by several scientific institutions and selected experts, and comments were incorporated by the coordinator and author prior to submission for final evaluation by a review group. The review group then took a decision as to the health risk assessment and proposed a guideline value.

During the preparation of draft evaluation documents and at the review group meetings, careful consideration was always given to previous risk assessments carried out by IPCS, in its Environmental Health Criteria monographs, the International Agency for Research on Cancer, the Joint FAO/WHO Meetings on Pesticide Residues, and the Joint FAO/WHO Expert Committee on Food Additives, which evaluates contaminants such as lead and cadmium in addition to food additives.

It is clear that not all the chemicals that may be found in drinking-water were evaluated in developing these guidelines. Chemicals of importance to Member States which have not been evaluated should be brought to the attention of WHO for inclusion in any future revision.

It is planned to establish a continuing process of revision of the *Guidelines for drinking-water quality* with a number of substances or agents subject to evaluation each year. Where appropriate, addenda will be issued, containing evaluations of new substances or substances already evaluated for which new scientific information has become available. Substances for which provisional guideline values have been established will receive high priority for re-evaluation.

# Acknowledgements

The work of the following coordinators was crucial in the development of Volumes 1 and 2 of the *Guidelines:*

J. K. Fawell, Water Research Centre, England (inorganic constituents)
J. R. Hickman, Department of National Health and Welfare, Canada (radioactive materials)
U. Lund, Water Quality Institute, Denmark (organic constituents and pesticides)
B. Mintz, Environmental Protection Agency, United States of America (disinfectants and disinfectant by-products)
E. B. Pike, Water Research Centre, England (microbiology)

The coordinator for Volume 3 of the *Guidelines* was J. Bartram of the Robens Institue of Health and Safety, England.

The WHO coordinators were as follows:

*Headquarters:* H. Galal-Gorchev, International Programme on Chemical Safety; R. Helmer, Division of Environmental Health.
*Regional Office for Europe:* X. Bonnefoy, Environment and Health; O. Espinoza, Environment and Health.

Ms Marla Sheffer of Ottawa, Canada, was responsible for the scientific editing of the guidelines.

The convening of the coordination and review group meetings was made possible by the financial support afforded to WHO by the Danish International Development Agency (DANIDA) and the following sponsoring countries: Belgium, Canada, France, Italy, Netherlands, United Kingdom of Great Britain and Northern Ireland and United States of America.

In addition, financial contributions for the convening of the final task group meeting were received from the Norwegian Agency for Development Cooperation (NORAD), the United Kingdom Overseas Development Administration (ODA) and the Water Services Association in the United Kingdom, the Swedish International Development Authority (SIDA), and the Government of Japan.

The efforts of all who helped in the preparation and finalization of the *Guidelines for drinking-water quality* are gratefully acknowledged.

# Acronyms and abbreviations used in the text

| | |
|---|---|
| ADI | acceptable daily intake |
| FAO | Food and Agriculture Organization of the United Nations |
| IARC | International Agency for Research on Cancer |
| ICRP | International Commission on Radiological Protection |
| ILO | International Labour Organisation |
| IPCS | International Programme on Chemical Safety |
| IQ | intelligence quotient |
| ISO | International Organization for Standardization |
| JECFA | Joint FAO/WHO Expert Committee on Food Additives |
| JMPR | Joint FAO/WHO Meeting on Pesticide Residues |
| LOAEL | lowest-observed-adverse-effect level |
| NOAEL | no-observed-adverse-effect level |
| NTU | nephelometric turbidity unit |
| PMTDI | provisional maximum tolerable daily intake |
| PTWI | provisional tolerable weekly intake |
| TCU | true colour unit |
| TDI | tolerable daily intake |
| UNEP | United Nations Environment Programme |
| WHO | World Health Organization. |

# 1.
# Introduction

This volume of the *Guidelines for drinking-water quality* explains how guideline values for drinking-water contaminants are to be used, defines the criteria used to select the various chemical, physical, microbiological, and radiological contaminants included in the report, describes the approaches used in deriving guideline values, and presents brief summary statements either supporting the guideline values recommended or explaining why no health-based guideline value is required at the present time.

This edition of the guidelines considers many drinking-water contaminants not included in the first edition. It also contains revised guideline values for many of the contaminants included in the first edition, which have been changed as a result of new scientific information. The guideline values given here supersede those in the 1984 edition.

Although the number of chemical contaminants for which guideline values are recommended is greater than in the first edition, it is unlikely that all of these chemical contaminants will occur in all water supplies or even in all countries. Care should therefore be taken in selecting substances for which national standards will be developed. A number of factors should be considered, including the geology of the region and the types of human activities that take place there. For example, if a particular pesticide is not used in the region, it is unlikely to occur in the drinking-water.

In other cases, such as the disinfection by-products, it may not be necessary to set standards for all of the substances for which guideline values have been proposed. If chlorination is practised, the trihalomethanes, of which chloroform is the major component, are likely to be the main disinfection by-products, together with the chlorinated acetic acids in some instances. In many cases, control of chloroform levels and, where appropriate, trichloroacetic acid will also provide an adequate measure of control over other chlorination by-products.

In developing national standards, care should also be taken to ensure that scarce resources are not unnecessarily diverted to the development of standards and the monitoring of substances of relatively minor importance.

Several of the inorganic elements for which guideline values have been recommended are recognized to be essential elements in human nutrition. No attempt has been made here to define a minimum desirable concentration of such substances in drinking-water.

## 1.1 General considerations

The primary aim of the *Guidelines for drinking-water quality* is the protection of public health. The guidelines are intended to be used as a basis for the development of national standards that, if properly implemented, will ensure the safety of drinking-water supplies through the elimination, or reduction to a minimum concentration, of constituents of water that are known to be hazardous to health. It must be emphasized that the guideline values recommended are not mandatory limits. In order to define such limits, it is necessary to consider the guideline values in the context of local or national environmental, social, economic, and cultural conditions.

The main reason for not promoting the adoption of international standards for drinking-water quality is the advantage provided by the use of a risk-benefit approach (qualitative or quantitative) to the establishment of national standards and regulations. This approach should lead to standards and regulations that can be readily implemented and enforced. For example, the adoption of drinking-water standards that are too stringent could limit the availability of water supplies that meet those standards – a significant consideration in regions of water shortage. The standards that individual countries will develop can thus be influenced by national priorities and economic factors. However, considerations of policy and convenience must never be allowed to endanger public health, and the implementation of standards and regulations will require suitable facilities and expertise as well as the appropriate legislative framework.

The judgement of safety – or what is an acceptable level of risk in particular circumstances – is a matter in which society as a whole has a role to play. The final judgement as to whether the benefit resulting from the adoption of any of the guideline values given here as standards justifies the cost is for each country to decide. What must be emphasized is that the guideline values have a degree of flexibility and enable a judgement to be made regarding the provision of drinking-water of acceptable quality.

Water is essential to sustain life, and a satisfactory supply must be made available to consumers. Every effort should be made to achieve a drinking-water quality as high as practicable. Protection of water supplies from contamination is the first line of defence. Source protection is almost invariably the best method of ensuring safe drinking-water and is to be preferred to treating a contaminated water supply to render it suitable for consumption. Once a potentially hazardous situation has been recognized, however, the risk to health, the availability of alternative sources, and the availability of suitable remedial measures must be considered so that a decision can be made about the acceptability of the supply.

As far as possible, water sources must be protected from contamination by human and animal waste, which can contain a variety of bacterial, viral, and protozoan pathogens and helminth parasites. Failure to provide adequate protection and effective treatment will expose the community to the risk of outbreaks of intestinal and other infectious diseases. Those at greatest risk of waterborne disease are infants and young children, people who are debilitated or living under unsanitary conditions, the sick, and the elderly. For these people, infective doses are significantly lower than for the general adult population.

The potential consequences of microbial contamination are such that its control must always be of paramount importance and must never be compromised.

The assessment of the risks associated with variations in microbial quality is difficult and controversial because of insufficient epidemiological evidence, the number of factors involved, and the changing interrelationships between these factors. In general terms, the greatest microbial risks are associated with ingestion of water that is contaminated with human and animal excreta. Microbial risk can never be entirely eliminated, because the diseases that are waterborne may also be transmitted by person-to-person contact, aerosols, and food intake; thus, a reservoir of cases and carriers is maintained. Provision of a safe water supply in these circumstances will reduce the chances of spread by these other routes. Waterborne outbreaks are particularly to be avoided because of their capacity to result in the simultaneous infection of a high proportion of the community.

The health risk due to toxic chemicals in drinking-water differs from that caused by microbiological contaminants. There are few chemical constituents of water that can lead to acute health problems except through massive accidental contamination of a supply. Moreover, experience shows that, in such incidents, the water usually becomes undrinkable owing to unacceptable taste, odour, and appearance.

The fact that chemical contaminants are not normally associated with acute effects places them in a lower priority category than microbial contaminants, the effects of which are usually acute and widespread. Indeed, it can be argued that chemical standards for drinking-water are of secondary consideration in a supply subject to severe bacterial contamination.

The problems associated with chemical constituents of drinking-water arise primarily from their ability to cause adverse health effects after prolonged periods of exposure; of particular concern are contaminants that have cumulative toxic properties, such as heavy metals, and substances that are carcinogenic.

It should be noted that the use of chemical disinfectants in water treatment usually results in the formation of chemical by-products, some of which are potentially hazardous. However, the risks to health from these by-products are extremely small in comparison with the risks associated with inadequate disinfection, and it is important that disinfection should not be compromised in attempting to control such by-products.

The radiological health risk associated with the presence of naturally occurring radionuclides in drinking-water should also be taken into consideration, although the contribution of drinking-water to total ambient exposure to these radionuclides is very small under normal circumstances. The guideline values recommended in this volume do not apply to water supplies contaminated during emergencies arising from accidental releases of radioactive substances to the environment.

In assessing the quality of drinking-water, the consumer relies principally upon his or her senses. Water constituents may affect the appearance, odour, or taste of the water, and the consumer will evaluate the quality and acceptability of the water on the basis of these criteria. Water that is highly turbid, is highly coloured, or has an objectionable taste or odour may be regarded by consumers as unsafe and may be rejected for drinking purposes. It is therefore vital to maintain a quality of water that is acceptable to the consumer, although the absence of any adverse sensory effects does not guarantee the safety of the water.

Countries developing national drinking-water limits or standards should carefully evaluate the costs and benefits associated with the control of aesthetic and organoleptic quality. Enforceable standards are sometimes set for contaminants directly related to health, whereas recommendations only are made for aesthetic and organoleptic characteristics. For countries with severely limited resources, it is even more important to establish priorities, and this should be done by considering the impact on health in each case. This approach does not underestimate the importance of the aesthetic quality of drinking-water. Source water that is aesthetically unsatisfactory may discourage the consumer from using an otherwise safe supply. Furthermore, taste, odour, and colour may be the first indication of potential health hazards.

Many parameters must be taken into consideration in the assessment of water quality, such as source protection, treatment efficiency and reliability, and protection of the distribution network (e.g., corrosion control). The costs associated with water quality surveillance and control must also be carefully evaluated before developing national standards. For guidance on these issues, the reader should refer to other more comprehensive publications (see Bibliography, page 144).

## 1.2 The nature of the guideline values

Guideline values have been set for potentially hazardous water constituents and provide a basis for assessing drinking-water quality.

# 1. INTRODUCTION

(*a*) A guideline value represents the concentration of a constituent that does not result in any significant risk to the health of the consumer over a lifetime of consumption.

(*b*) The quality of water defined by the *Guidelines for drinking-water quality* is such that it is suitable for human consumption and for all usual domestic purposes, including personal hygiene. However, water of a higher quality may be required for some special purposes, such as renal dialysis.

(*c*) When a guideline value is exceeded, this should be a signal: (i) to investigate the cause with a view to taking remedial action; (ii) to consult with, and seek advice from, the authority responsible for public health.

(*d*) Although the guideline values describe a quality of water that is acceptable for lifelong consumption, the establishment of these guideline values should not be regarded as implying that the quality of drinking-water may be degraded to the recommended level. Indeed, a continuous effort should be made to maintain drinking-water quality at the highest possible level.

(*e*) Short-term deviations above the guideline values do not necessarily mean that the water is unsuitable for consumption. The amount by which, and the period for which, any guideline value can be exceeded without affecting public health depends upon the specific substance involved.

It is recommended that when a guideline value is exceeded, the surveillance agency (usually the authority responsible for public health) should be consulted for advice on suitable action, taking into account the intake of the substance from sources other than drinking-water (for chemical constituents), the toxicity of the substance, the likelihood and nature of any adverse effects, the practicability of remedial measures, and similar factors.

(*f*) In developing national drinking-water standards based on these guideline values, it will be necessary to take account of a variety of geographical, socioeconomic, dietary, and other conditions affecting potential exposure. This may lead to national standards that differ appreciably from the guideline values.

(*g*) In the case of radioactive substances, screening values for gross alpha and gross beta activity are given, based on a reference level of dose.

It is important that recommended guideline values are both practical and feasible to implement as well as protective of public health. Guideline values are not set at concentrations lower than the detection limits achievable under routine laboratory operating conditions. Moreover, guideline values are recommended only when control techniques are available to remove or reduce the concentration of the contaminant to the desired level.

In some instances, *provisional* guideline values have been set for constituents for which there is some evidence of a potential hazard but where the available information on health effects is limited. Provisional guideline values have also been set for substances for which the calculated guideline value would be (i) below the practical quantification level, or (ii) below the level that can be achieved through practical treatment methods. Finally, provisional guideline values have been set for certain substances when it is likely that guideline values will be exceeded as a result of disinfection procedures.

Aesthetic and organoleptic characteristics are subject to individual preference as well as social, economic, and cultural considerations. For this reason, although guidance can be given on the levels of substances that may be aesthetically unacceptable, no guideline values have been set for such substances where they do not represent a potential hazard to health.

The recommended guideline values are set at a level to protect human health; they may not be suitable for the protection of aquatic life. The guidelines apply to bottled water and ice intended for human consumption but do not apply to natural mineral waters, which should be regarded as beverages rather than drinking-water in the usual sense of the word. The Codex Alimentarius Commission has developed Codex standards for such mineral waters.

## 1.3 Criteria for the selection of health-related drinking-water contaminants

The recognition that faecally polluted water can lead to the spread of microbial infections has led to the development of sensitive methods for routine examination to ensure that water intended for human consumption is free from faecal contamination. Although it is now possible to detect the presence of many pathogens in water, the methods of isolation and enumeration are often complex and time-consuming. It is therefore impracticable to monitor drinking-water for every possible microbial pathogen. A more logical approach is the detection of organisms normally present in the faeces of humans and other warm-blooded animals as indicators of faecal pollution, as well as of the efficacy of water treatment and disinfection. The various bacterial indicators used for this purpose are described in section 2.2. The presence of such organisms indicates the presence of faecal material and, hence, that intestinal pathogens could be present. Conversely, their absence indicates that pathogens are probably also absent.

Thousands of organic and inorganic chemicals have been identified in drinking-water supplies around the world, many in extremely low concentrations. The chemicals selected for the development of guideline values include those considered potentially hazardous to human health, those detected relatively frequently in drinking-water, and those detected in relatively high concentrations.

Some potentially hazardous chemicals in drinking-water are derived directly from treatment chemicals or construction materials used in water supply systems.

Such chemicals are best controlled by appropriate specifications for the chemicals and materials used. For example, a wide range of polyelectrolytes are now used as coagulant aids in water treatment, and the presence of residues of the unreacted monomer may cause concern. Many polyelectrolytes are based on acrylamide polymers and co-polymers, in both of which the acrylamide monomer is present as a trace impurity. Chlorine used for disinfection has sometimes been found to contain carbon tetrachloride. This type of drinking-water contamination is best controlled by the application of regulations governing the quality of the products themselves rather than the quality of the water. Similarly, strict national regulations on the quality of pipe material should avoid the possible contamination of drinking-water by trace constituents of plastic pipes. The control of contamination of water supplies by *in situ* polymerized coatings and coatings applied in a solvent requires the development of suitable codes of practice, in addition to controls on the quality of the materials used.

# 2.
# Microbiological aspects

## 2.1 Agents of significance

### 2.1.1 Waterborne infections

> Infectious diseases caused by pathogenic bacteria, viruses, and protozoa or by parasites are the most common and widespread health risk associated with drinking-water.

Infectious diseases are transmitted primarily through human and animal excreta, particularly faeces. If there are active cases or carriers in the community, then faecal contamination of water sources will result in the causative organisms being present in the water. The use of such water for drinking or for preparing food, contact during washing or bathing, and even inhalation of water vapour or aerosols may then result in infection.

### 2.1.2 Orally transmitted infections of high priority

The human pathogens that can be transmitted orally by drinking-water are listed in Table 1 (p. 10), together with a summary of their health significance and main properties. Those that present a serious risk of disease whenever present in drinking-water include *Salmonella* spp., *Shigella* spp., pathogenic *Escherichia coli*, *Vibrio cholerae*, *Yersinia enterocolitica*, *Campylobacter jejuni*, and *Campylobacter coli*, the viruses listed in Table 1, and the parasites *Giardia* spp., *Cryptosporidium* spp., *Entamoeba histolytica*, and *Dracunculus medinensis*. Most of these pathogens are distributed worldwide. However, outbreaks of cholera and infection by the guinea worm *D. medinensis* are regional. The elimination of all these agents from water intended for drinking has high priority. Eradication of *D. medinensis* is a recognized target of the World Health Assembly (World Health Assembly resolution WHA44.5, 1991).

## 2.1.3 Opportunistic and other water-associated pathogens

Other pathogens are accorded moderate priority in Table 1 or are not listed, either because they are of low pathogenicity, causing disease opportunistically in subjects with low or impaired immunity, or because, even though they cause serious diseases, the primary route of infection is by contact or inhalation, rather than by ingestion.

Opportunistic pathogens are naturally present in the environment and are not formally regarded as pathogens. They are able to cause disease in people with impaired local or general defence mechanisms, such as the elderly or the very young, patients with burns or extensive wounds, those undergoing immunosuppressive therapy, or those with acquired immunodeficiency syndrome (AIDS). Water used by such patients for drinking or bathing, if it contains large numbers of these organisms, can produce various infections of the skin and the mucous membranes of the eye, ear, nose, and throat. Examples of such agents are *Pseudomonas aeruginosa* and species of *Flavobacterium, Acinetobacter, Klebsiella, Serratia, Aeromonas,* and certain "slow-growing" mycobacteria.

Certain serious illnesses result from inhalation of water in which the causative organisms have multiplied because of warm temperatures and the presence of nutrients. These include Legionnaires' disease (*Legionella* spp.) and those caused by the amoebae *Naegleria fowleri* (primary amoebic meningoencephalitis) and *Acanthamoeba* spp. (amoebic meningitis, pulmonary infections).

Schistosomiasis (bilharziasis) is a major parasitic disease of tropical and subtropical regions, and is primarily spread by contact with water during bathing or washing. The larval stage (cercariae) released by infected aquatic snails penetrates the skin. If pure drinking-water is readily available, it will be used for washing, and this will have the benefit of reducing the need to use contaminated surface water.

It is conceivable that unsafe drinking-water contaminated with soil or faeces could act as a carrier of other parasitic infections, such as balantidiasis (*Balantidium coli*), and certain helminths (species of *Fasciola, Fasciolopsis, Echinococcus, Spirometra, Ascaris, Trichuris, Toxocara, Necator, Ancylostoma, Strongyloides* and *Taenia solium*). However, in most of these, the normal mode of transmission is ingestion of the eggs in food contaminated with faeces or faecally contaminated soil (in the case of *Taenia solium*, ingestion of the larval cysticercus stage in uncooked pork) rather than ingestion of contaminated drinking-water.

## 2.1.4 Toxins from *Cyanobacteria*

Blooms of *Cyanobacteria* (commonly called blue-green algae) occur in lakes and reservoirs used for potable supply. Three types of toxin can be produced, depending upon species:

## Table 1. Orally transmitted waterborne pathogens and their significance in water supplies

| Pathogen | Health significance | Persistence in water supplies[a] | Resistance to chlorine[b] | Relative infective dose[c] | Important animal reservoir |
|---|---|---|---|---|---|
| **Bacteria** | | | | | |
| Campylobacter jejuni, C. coli | High | Moderate | Low | Moderate | Yes |
| Pathogenic Escherichia coli | High | Moderate | Low | High | Yes |
| Salmonella typhi | High | Moderate | Low | High[d] | No |
| Other salmonellae | High | Long | Low | High | Yes |
| Shigella spp. | High | Short | Low | Moderate | No |
| Vibrio cholerae | High | Short | Low | High | No |
| Yersinia enterocolitica | High | Long | Low | High(?) | Yes |
| Pseudomonas aeruginosa[e] | Moderate | May multiply | Moderate | High(?) | No |
| Aeromonas spp. | Moderate | May multiply | Low | High(?) | No |
| **Viruses** | | | | | |
| Adenoviruses | High | ? | Moderate | Low | No |
| Enteroviruses | High | Long | Moderate | Low | No |
| Hepatitis A | High | ? | Moderate | Low | No |
| Enterically transmitted non-A, non-B hepatitis viruses, hepatitis E | High | ? | ? | Low | No |
| Norwalk virus | High | ? | ? | Low | No |
| Rotavirus | High | ? | ? | Moderate | No(?) |
| Small round viruses | Moderate | ? | ? | Low(?) | No |

? — not known or uncertain.

[a] Detection period for infective stage in water at 20 °C : short, up to 1 week; moderate, 1 week to 1 month; long, over 1 month.

[b] When the infective stage is freely suspended in water treated at conventional doses and contact times. Resistance moderate, agent may not be completely destroyed; resistance low, agent completely destroyed.

[c] Dose required to cause infection in 50% of healthy adult volunteers; may be as little as one infective unit for some viruses.

[d] From experiments with human volunteers (see section 2.1.7)

[e] Main route of infection is by skin contact, but can infect immunosuppressed or cancer patients orally.

## 2. MICROBIOLOGICAL ASPECTS

*Table 1 (continued)*

| Pathogen | Health significance | Persistence in water supplies[a] | Resistance to chlorine[b] | Relative infective dose[c] | Important animal reservoir |
|---|---|---|---|---|---|
| **Protozoa** | | | | | |
| Entamoeba histolytica | High | Moderate | High | Low | No |
| Giardia intestinalis | High | Moderate | High | Low | Yes |
| Cryptosporidium parvum | High | Long | High | Low | Yes |
| **Helminths** | | | | | |
| Dracunculus medinensis | High | Moderate | Moderate | Low | Yes |

— hepatotoxins, produced by species of *Microcystis, Oscillatoria, Anabaena,* and *Nodularia,* typified by microcystin LR:R, which induce death by circulatory shock and massive liver haemorrhage within 24 hours of ingestion;
— neurotoxins, produced by species of *Anabaena, Oscillatoria, Nostoc, Cylindrospermum,* and *Aphanizomenon;*
— lipopolysaccharides.

There are a number of unconfirmed reports of adverse health effects caused by algal toxins in drinking-water, including an epidemiological study of mild, reversible liver damage in hospital patients receiving drinking-water from a reservoir with a very large toxic bloom of *Microcystis aeruginosa.* Only activated carbon and ozonation appear to remove or reduce toxicity; however, knowledge is impeded by the lack of suitable analytical methods. There are insufficient data to allow guidelines to be recommended, but the need to protect impounded surface water sources from discharges of nutrient-rich effluents is emphasized.

### 2.1.5 Nuisance organisms

There are a number of diverse organisms that have no public health significance but which are undesirable because they produce turbidity, taste and odour, or because they appear as visible animal life in water. As well as being aesthetically objectionable, they indicate that water treatment and the state of maintenance and repair of the system are defective. Examples include:

- seasonal blooms of cyanobacteria and other algae in reservoirs and in river waters, impeding coagulation and filtration and causing coloration and turbidity of water after filtration;
- in waters containing ferrous and manganous salts, oxidation by iron bacteria, causing rust-coloured deposits on the walls of tanks, pipes and channels, and carry-over of deposits in the water;
- microbial corrosion of iron and steel pipes by iron and sulfur bacteria;
- production of objectionable tastes and odours, with a low threshold, e.g., geosmin and 2-methylisoborneol by actinomycetes and cyanobacteria;
- colonization of unsuitable non-metallic fittings, pipes, jointing compounds and lining materials by microorganisms able to utilize leached organic compounds;
- microbial growth in distribution systems encouraged by the presence of biodegradable and assimilable organic carbon in water, often released by oxidative disinfectants (chlorine, ozone); this growth may include *Aeromonas* spp., which can produce false positive reactions in the coliform test;
- infestation of water mains by animal life, feeding on microbial growth in the water or on slimes, for example crustacea (*Gammarus pulex, Crangonyx pseudogracilis, Cyclops* spp., and *Chydorus sphaericus*), *Asellus aquaticus*, snails, mussels (*Dreissena polymorpha*), bryozoa (*Plumatella*), *Nais* worms, nematodes, and larvae of chironomids (*Chironomus* spp.) and mosquitos (*Culex* spp.); in warm weather, slow sand filters can sometimes discharge chironomid larvae by draw-down into the filtered water.

The only positively identified health hazard from animal life in drinking-water arises with the intermediate stage of the guinea worm, *Dracunculus medinensis*, which parasitizes the water flea, *Cyclops*.

## 2.1.6 Persistence in water

After leaving the body of their host, pathogens and parasites gradually lose viability and the ability to infect. The rate of decay is usually exponential, and a pathogen will become undetectable after a certain period. Pathogens with low persistence must rapidly find a new host and are more likely to be spread by person-to-person contact or faulty personal or food hygiene than by drinking-water. Because faecal contamination is usually dispersed rapidly in surface waters, the most common waterborne pathogens and parasites are those that have high infectivity or possess high resistance to decay outside the body. Persistence in water and resistance to chlorination are summarized in Table 1, page 10.

Persistence is affected by several factors, of which temperature is the most important. Decay is usually accelerated by increasing temperature of water and may be mediated by the lethal effects of ultraviolet radiation in sunlight acting near the water surface. Viruses and the resting stages of parasites (cysts, oocysts, ova) are unable to multiply in water. Conversely, relatively high amounts of biodegradable organic carbon, together with warm temperatures and low residual concentrations of chlorine, can permit growth of *Legionella, Naegleria fowleri, Acanthamoeba,* the opportunistic pathogens *Pseudomonas aeruginosa* and *Aeromonas,* and nuisance organisms during water distribution.

### 2.1.7 Infective dose

Waterborne transmission of the pathogens listed in Table 1 has been confirmed by epidemiological studies and case histories. Part of the demonstration of pathogenicity involves reproducing the disease in suitable hosts. Experimental studies of infectivity provide relative information, as shown in Table 1, but it is doubtful whether the infective doses obtained are relevant to natural infections. For example, many epidemics of typhoid fever can be explained only by assuming that the infective dose was very low. Individuals vary widely in immunity, whether acquired by contact with a pathogen or influenced by such factors as age, sex, state of health, and living conditions. Pathogens are likely to be widely dispersed and diluted in drinking-water, and a large number of people will be exposed to relatively small numbers. Hence, the minimal infective doses and the attack rates are likely to be lower than in experimental studies. If food is contaminated by water containing pathogens that multiply subsequently, or if a susceptible person becomes infected by water, subsequently infecting others by person-to-person contact, the initial involvement of water may be unsuspected. Hence, improvements in water supply, sanitation, and hygiene are closely linked in control of disease in a community.

The multifactorial natures of infection and immunity mean that experimental data from infectivity studies and epidemiology cannot by used to predict infective doses or risk precisely. However, probabilistic modelling has been used to predict the effects of water treatment in reducing attack rates from very low doses of viruses and *Giardia* and thereby to confirm water treatment criteria.

### 2.1.8 Guideline values

Pathogenic agents have several properties that distinguish them from chemical pollutants:

- Pathogens are discrete and not in solution.
- Pathogens are often clumped or adherent to suspended solids in water, so that the likelihood of acquiring an infective dose cannot be predicted from their average concentration in water.

- The likelihood of a successful challenge by a pathogen, resulting in infection, depends upon the invasiveness and virulence of the pathogen, as well as upon the immunity of the individual.
- If infection is established, pathogens multiply in their host. Certain pathogenic bacteria are also able to multiply in food or beverages, thereby perpetuating or even increasing the chances of infection.
- Unlike many chemical agents, the dose response of pathogens is not cumulative.

Because of these properties there is **no tolerable lower limit** for pathogens, and water intended for consumption, for preparing food and drink, or for personal hygiene should thus contain no agents pathogenic for humans. Pathogen-free water is attainable by selection of high-quality uncontaminated sources of water, by efficient treatment and disinfection of water known to be contaminated with human or animal faeces, and by ensuring that such water remains free from contamination during distribution to the user. Such a policy creates multiple barriers to the transmission of infection (see Chapter 6 for a more detailed discussion of the multiple-barrier concept).

As indicated in section 1.3, although many pathogens can be detected by suitable methods, it is easier to test for bacteria that specifically indicate the presence of faecal pollution or the efficiency of water treatment and disinfection (see section 2.2). It follows that water intended for human consumption should contain none of these bacteria. In the great majority of cases, monitoring for indicator bacteria provides a great factor of safety because of their large numbers in polluted waters; this has been reinforced over many years of experience.

## 2.2 Microbial indicators of water quality

### 2.2.1 Introduction

Frequent examinations for faecal indicator organisms remain the most sensitive and specific way of assessing the hygienic quality of water. Faecal indicator bacteria should fulfil certain criteria to give meaningful results. They should be universally present in high numbers in the faeces of humans and warm-blooded animals, and readily detectable by simple methods, and they should not grow in natural water. Furthermore, it is essential that their persistence in water and their degree of removal in treatment of water are similar to those of waterborne pathogens. The major indicator organisms of faecal pollution – *Escherichia coli*, the thermotolerant and other coliform bacteria, the faecal streptococci, and spores of sulfite-reducing clostridia – are described briefly below. Details of additional microbial indicators of water quality, such as heterotrophic plate-count bacteria, bacteriophages, and opportunistic and overt pathogens, are given in Volume 2 of *Guidelines for drinking-water quality*.

## 2.2.2 General principles

While the criteria described above for an ideal faecal indicator are not all met by any one organism, many of them are fulfilled by *E. coli* and, to a lesser extent, by the thermotolerant coliform bacteria. The faecal streptococci satisfy some of the criteria, although not to the same extent as *E. coli*, and they can be used as supplementary indicators of faecal pollution or treatment efficiency in certain circumstances. It is recommended that *E. coli* is the indicator of first choice when resources for microbiological examination are limited. Because enteroviruses and the resting stages of *Cryptosporidium, Giardia*, amoebae, and other parasites are known to be more resistant to disinfection than *E. coli* and faecal streptococci, the absence of the latter organisms will not necessarily indicate freedom from the former. Spores of sulfite-reducing clostridia can be used as an additional parameter in this respect.

## 2.2.3 *Escherichia coli* and the coliform bacteria

### *Escherichia coli*

*Escherichia coli* is a member of the family Enterobacteriaceae, and is characterized by possession of the enzymes $\beta$-galactosidase and $\beta$-glucuronidase. It grows at 44–45 °C on complex media, ferments lactose and mannitol with the production of acid and gas, and produces indole from tryptophan. Some strains can grow at 37 °C, but not at 44–45 °C, and some do not produce gas. *E. coli* does not produce oxidase or hydrolyse urea. Complete identification of *E. coli* is too complicated for routine use, hence certain tests have been evolved for identifying the organism rapidly with a high degree of certainty. Some of these methods have been standardized at international and national levels and accepted for routine use, whereas others are still in the developmental or evaluative stage.

*E. coli* is abundant in human and animal faeces, where it may attain concentrations in fresh faeces of $10^9$ per gram. It is found in sewage, treated effluents, and all natural waters and soils that are subject to recent faecal contamination, whether from humans, agriculture, or wild animals and birds. Recently, it has been suggested that *E. coli* may be found or even multiply in tropical waters that are not subject to human faecal pollution. However, even in the remotest regions, faecal contamination by wild animals, including birds, can never be excluded. As animals can transmit pathogens infective for humans, the presence of *E. coli* or thermotolerant coliform bacteria can never be ignored, because the presumptions remain that the water has been faecally contaminated and that treatment has been ineffective.

### Thermotolerant coliform bacteria

These are defined as the group of coliform organisms that are able to ferment lactose at 44–45 °C; they comprise the genus *Escherichia* and, to a lesser extent, species of *Klebsiella, Enterobacter,* and *Citrobacter.* Thermotolerant coliforms other than *E. coli* may also originate from organically enriched water such as industrial effluents or from decaying plant materials and soils. For this reason, the often-used term "faecal" coliforms is not correct, and its use should be discontinued.

Regrowth of thermotolerant coliform organisms in the distribution system is unlikely unless sufficient bacterial nutrients are present or unsuitable materials are in contact with the treated water, water temperature is above 13 °C, and there is no free residual chlorine.

The concentrations of thermotolerant coliforms are, under most circumstances, directly related to that of *E. coli*. Hence, their use in assessing water quality is considered acceptable for routine purposes. The limitations with regard to specificity should always be borne in mind when the data are interpreted. Specific detection of *E. coli* by additional confirmatory tests or by direct methods, as described in the research literature, should be carried out if high counts of thermotolerant coliforms are found in the absence of detectable sanitary hazards. National reference laboratories are advised to examine the specificity of the thermotolerant coliform test for *E. coli* under local circumstances when developing national standard methods.

Because thermotolerant coliform organisms are readily detected, they have an important secondary role as indicators of the efficiency of water treatment processes in removing faecal bacteria. They may therefore be used in assessing the degree of treatment necessary for waters of different quality and for defining targets of performance for bacterial removal (see section 2.3).

### Coliform organisms (total coliforms)

Coliform organisms have long been recognized as a suitable microbial indicator of drinking-water quality, largely because they are easy to detect and enumerate in water. The term "coliform organisms" refers to Gram-negative, rod-shaped bacteria capable of growth in the presence of bile salts or other surface-active agents with similar growth-inhibiting properties and able to ferment lactose at 35–37 °C with the production of acid, gas, and aldehyde within 24–48 hours. They are also oxidase-negative and non-spore-forming. By definition, coliform bacteria display $\beta$-galactosidase activity.

Traditionally, coliform bacteria were regarded as belonging to the genera *Escherichia, Citrobacter, Enterobacter,* and *Klebsiella*. However, as defined by modern taxonomical methods, the group is heterogeneous. It includes lactose-fermenting bacteria, such as *Enterobacter cloacae* and *Citrobacter freundii,* that can be found both in faeces and the environment (nutrient-rich waters, soil, decaying plant material), and also in drinking-water with relatively high concentrations of nutrients, as well as species that are rarely, if ever, found in faeces and may multiply in relatively good quality drinking-waters, for example, *Serratia fonticola, Rahnella aquatilis,* and *Buttiauxella agrestis.*

The existence both of non-faecal bacteria that fit the definitions of coliform bacteria and of lactose-negative coliform bacteria limits the applicability of this group as an indicator of faecal pollution. Coliform bacteria should not be detectable in treated water supplies and, if found, suggest inadequate treatment, post-treatment contamination, or excessive nutrients. The coliform test can therefore be used as an indicator of treatment efficiency and of the integrity of the distribution system. Although coliform organisms may not always be directly related to the presence of faecal contamination or pathogens in drinking-water, the coliform test is still useful for monitoring the microbial quality of treated piped water supplies. If there is any doubt, especially when coliform organisms are found in the absence of thermotolerant coliform organisms and *E. coli,* identification to the species level or analyses for other indicator organisms may be undertaken to investigate the nature of the contamination. Sanitary inspections will also be needed.

## 2.2.4 Faecal streptococci

The term "faecal streptococci" refers to those streptococci generally present in the faeces of humans and animals. All possess the Lancefield group D antigen. Taxonomically, they belong to the genera *Enterococcus* and *Streptococcus.* The taxonomy of enterococci has recently undergone important changes, and detailed knowledge of the ecology of many of the new species is lacking. The genus *Enterococcus* now includes all streptococci that share certain biochemical properties and have a wide tolerance of adverse growth conditions. It includes the species *E. avium, E. casseliflavus, E. cecorum, E. durans, E. faecalis, E. faecium, E. gallinarum, E. hirae, E. malodoratus, E. mundtii,* and *E. solitarius.* Most of these species are of faecal origin and can generally be regarded as specific indicators of human faecal pollution under many practical circumstances. They may, however, be isolated from the faeces of animals, and certain species and subspecies, such as *E. casseliflavus, E. faecalis* var. *liquefaciens, E. malodoratus,* and *E. solitarius,* occur primarily on plant material.

In the genus *Streptococcus,* only *S. bovis* and *S. equinus* possess the group D antigen and are members of the faecal streptococcus group. Their sources are mainly animal faeces. Faecal streptococci rarely multiply in polluted water, and they are more persistent than *E. coli* and coliform bacteria. Their primary value in water quality examination is therefore as additional indicators of treatment efficiency. Furthermore, streptococci are highly resistant to drying and may be valuable for routine control after laying new mains or repairs in distribution systems, or for detecting pollution by surface run-off to ground or surface waters.

## 2.2.5 Sulfite-reducing clostridia

These are anaerobic, spore-forming organisms, of which the most characteristic, *Clostridium perfringens* (*C. welchii*), is normally present in faeces, although in much smaller numbers than *E. coli*. However, they are not exclusively of faecal origin and can be derived from other environmental sources. Clostridial spores can survive in water much longer than organisms of the coliform group and will resist disinfection. Their presence in disinfected waters may thus indicate deficiencies in treatment and that disinfection-resistant pathogens could have survived treatment. In particular, the presence of *C. perfringens* in filtered supplies may indicate deficiencies in filtration practice. Because of their longevity, they are best regarded as indicating intermittent or remote contamination. They thus have a special value but are not recommended for routine monitoring of distribution systems. Because they tend to survive and accumulate, they may be detected long after and far from the pollution and thus give rise to false alarms.

## 2.2.6 Coliphages and other alternative indicators

The bacteriophages have been proposed as indicators of water quality because of their similarity to human enteroviruses and their easy detection in water. Two groups have been studied extensively: the somatic coliphages, which infect *E. coli* host strains through cell-wall receptors; and the F-specific RNA-bacteriophages, which infect strains of *E. coli* and related bacteria through the F- or sex-pili. Neither occurs in high numbers in fresh human or animal faeces, but they are abundant in sewage. Their significance is as indicators of sewage contamination and, because of their greater persistence compared with bacterial indicators, as additional indicators of treatment efficiency or groundwater protection.

The bifidobacteria and the *Bacteroides fragilis* group are very numerous in faeces but have not been considered as suitable indicators of faecal pollution (see Volume 2) because they decay more rapidly in water than coliform bacteria and because the methods of examination are not very reliable and have not been standardized.

## 2.2.7 Methods of detection

Microbiological examination provides the most sensitive, although not the most rapid, indication of pollution of drinking-water supplies. Unlike chemical or physical analysis, however, it is a search for very small numbers of viable organisms and not for a defined chemical entity or physical property. Because the growth medium and the conditions of incubation, as well as the nature and age of the water sample, can influence the species isolated and the count, microbiological examinations may have variable accuracy. This means that the standardization of methods and of laboratory procedures is of great importance if criteria for microbiological quality of water are to be uniform in different laboratories and internationally. International standard methods should be evaluated under local circumstances before being adopted in national surveillance programmes. Established standard methods are available, such as those of the International Organization for Standardization (ISO) (Table 2), of the American Public Health Association (APHA), and of the United Kingdom Department of Health and Social Security. It is desirable that established standard methods should be used for routine examinations. Whatever method is chosen for detection of *E. coli* and the coliform group, some step for "resuscitating" or recovering environmentally- or disinfectant-damaged strains must be used, such as pre-incubation for a short period at a lower temperature.

*Table 2. International Organization for Standardization (ISO) standards for detection and enumeration of faecal indicator bacteria in water*

| ISO standard no. | Title (water quality) |
| --- | --- |
| 6461-1:1986 | Detection and enumeration of the spores of sulfite-reducing anaerobes (clostridia) - Part 1: Method by enrichment in a liquid medium |
| 6461-2:1986 | Detection and enumeration of the spores of sulfite-reducing anaerobes (clostridia) - Part 2: Method by membrane filtration |
| 7704:1985 | Evaluation of membrane filters used for microbiological analyses |
| 7899-1:1984 | Detection and enumeration of faecal streptococci - Part 1: Method by enrichment in a liquid medium |
| 7899-2:1984 | Dectection and enumeration of faecal streptococci - Part 2: Method by membrane filtration |
| 9308-1:1990 | Detection and enumeration of coliform organisms, thermotolerant coliform organisms, and presumptive *Escherichia coli* - Part 1: Membrane filtration method |
| 9308-2:1990 | Dectection and enumeration of coliform organisms, thermotolerant coliform organisms, and presumptive *Escherichia coli* - Part 2: Multiple tube (most probable number) method. |

## 2.3 Recommendations

### 2.3.1 General principles

The provision of a safe supply of drinking-water depends upon use of either a protected high-quality ground water or a properly selected and operated series of treatments capable of reducing pathogens and other contaminants to negligible levels, not injurious to health. Treatment systems should provide multiple barriers to the transmission of infection. The processes preceding terminal disinfection should be capable of producing water of high microbiological quality, so that terminal disinfection becomes a final safeguard. Disinfection is also most efficient when the water has already been treated to remove turbidity and when substances exerting a disinfectant demand, or capable of protecting pathogens from disinfection, have been removed as far as possible.

The search for microbial indicators of faecal pollution is a "fail-safe" concept; in other words, if faecal indicators are shown to be present, then it must be assumed that pathogens could also be present. For this reason, faecal indicator bacteria must never be present in treated water delivered to the consumer, and any detection should prompt immediate action to discover the cause and to take remedial action.

The most specific of the readily detectable faecal indicator bacteria and the one present in greatest numbers in faeces is *Escherichia coli* and it is therefore recommended as the indicator of choice for drinking-water. The thermotolerant coliform test can be used as an alternative to the test for *E. coli*. Thermotolerant coliform bacteria are also recommended as indicators of the efficiency of water treatment processes in removing enteric pathogens and faecal bacteria, and for grading the quality of source waters in order to select the intensity of treatment needed. Total coliform bacteria should not be present in treated water supplies and, if found, suggest inadequate treatment, post-treatment contamination, or excessive nutrients.

### 2.3.2 Selection of treatment processes

The selection of treatment processes to meet microbiological and chemical requirements can be made only after a careful detailed survey of the source and watershed, as outlined in section 6.2, including assessment of likely sources of pollution. Extensive bacteriological surveys, to include different seasons and weather conditions, can be used to assist in the selection. Regular bacteriological examination of source water after commissioning the treatment plant will establish long-term trends in quality and indicate whether there is a need to revise the treatment given.

### 2.3.3 Treatment objectives

The multiple-barrier concept of water treatment (see Chapter 6) requires that the removal of pathogens and of pollutants and biodegradable compounds should be as nearly complete as possible before terminal disinfection. Table 3 gives an example of performance objectives for typical urban water treatment processes, based upon loadings and removal of turbidity and thermotolerant coliform bacteria. These levels of performance are capable of being met and exceeded comfortably in normal operation. It is emphasized that the sequence of processes given in Table 3 is only one example from the many possible combinations of processes that are used in normal practice.

*Table 3. An example to illustrate the level of performance that can be achieved in removal of turbidity and thermotolerant coliform bacteria in conventional urban water treatment*

| Stage and process | Turbidity | | | Thermotolerant coliform bacteria | | |
|---|---|---|---|---|---|---|
| | Removal[a] (%) | Average loading (NTU)[b] | Maximum loading (NTU)[b] | Removal[a] (%) | Average loading (per 100 ml) | Maximum loading (per 100 ml) |
| Micro-straining | NA[c] | NA | NA | NA | NA | NA |
| Pretreatment[d] | NA | NA | NA | >99.9 | 1000 | 10 000 |
| Coagulation/settling[e] | 90 | 50 | 300 | NA | NA | NA |
| Rapid filtration[e] | >80 | 5 | 30 | 80 | 1 | 10 |
| Terminal chlorination | NA | 1 | 5 | >99.9 | <1 | 2 |
| Mains distribution | NA | <1 | <5 | NA | <1 | <1 |

[a] Required performance.
[b] NTU, nephelometric turbidity units.
[c] NA, not applicable. Process not designed to remove turbidity and/or bacteria. Micro-straining removes micro-algae and zooplankton.
[d] Pretreatments that can result in significant reductions in thermotolerant coliform bacteria are storage in reservoirs for 3–4 weeks, and pre-disinfection.
[e] Taken together, coagulation, settling, and rapid filtration should be expected to remove 99.9% of thermotolerant coliform bacteria.

The multiple-barrier concept can also be applied to water treatment in rural and remote regions. Table 4 gives an example of treatment objectives for such plants.

### 2.3.4 Guideline values

It is most important that the reasons for adopting the following guideline values for drinking-water are properly understood and that the guideline values are used only in conjunction with the information given below and in Volume 2.

**Table 4. An example of performance objectives for removal of turbidity and thermotolerant coliform bacteria in small-scale water treatment**

| Stage and process | Turbidity | | | Thermotolerant coliform bacteria | | |
|---|---|---|---|---|---|---|
| | Removal[a] (%) | Average loading (NTU)[b] | Maximum loading (NTU)[b] | Removal[a] (%) | Average loading (per 100 ml) | Maximum loading (per 100 ml) |
| Screening | NA[c] | NA | NA | NA | NA | NA |
| Plain sedimentation | 50 | 60 | 600 | 50 | 1000 | 10 000 |
| Gravel pre-filters (3-stage) | 80 | 30 | 300 | 90 | 500 | 5000 |
| Slow sand filter | >90 | 6 | 60 | 95 | 50 | 500 |
| Disinfection | NA | <1 | <5 | >99.9 | <3 | 25 |
| Distributed water | NA | <1 | <5 | NA | <1 | <1 |

[a] Required performance.
[b] NTU, nephelometric turbidity units.
[c] NA, not applicable. Process not designed to remove turbidity and/or bacteria.

## Bacteriological quality

Water intended for drinking and household purposes must not contain waterborne pathogens. Because the most numerous and the most specific bacterial indicator of faecal pollution from humans and animals is *E. coli*, it follows that *E. coli* or thermotolerant coliform organisms must not be present in 100-ml samples of any water intended for drinking (see Annex 2, Table A2.1).

This criterion is readily achievable by water treatment (see section 6.3). In nearly all epidemics of waterborne disease, it has been shown that the bacteriological quality of the water was unsatisfactory and that there was evidence of failure in terminal disinfection.

During distribution, the bacteriological quality of water may deteriorate. Coliform bacteria other than *E. coli* can occur in inadequately treated supplies, or those contaminated after leaving the treatment plant, as a result of growth in sediments and on unsuitable materials in contact with the water (washers, packing, lubricants, plastics and plasticizers, for example). They may also gain entrance from soil or natural water through leaky valves and glands, repaired mains, or back-siphonage. This type of contamination is most likely to be found when the water is untreated or undisinfected, or where there is limited or no residual disinfectant. Allowance can be made for the occasional occurrence in the distribution system of coliform organisms in up to 5% of samples taken over any 12-month period, provided *E. coli* is not present (Table A2.1, p. 173). It must be stressed that any regular occurrence of coliform organisms is a matter of concern and should be investigated.

## *Virological quality*

Drinking-water must essentially be free of human enteroviruses to ensure negligible risk of transmitting viral infection. Any drinking-water supply subject to faecal contamination presents a risk of viral disease to consumers. Two approaches can be used to ensure that the risk of viral infection is kept to a minimum: providing drinking-water from a source verified free of faecal contamination, or adequately treating faecally contaminated water to reduce enteroviruses to a negligible level.

Virological studies have shown that drinking-water treatment can considerably reduce the levels of viruses but may not eliminate them completely from very large volumes of water. Virological, epidemiological, and risk analyses are providing important information, although it is still insufficient for deriving quantitative and direct virological criteria. Such criteria cannot be recommended for routine use because of the cost, complexity, and lengthy nature of virological analyses, and the fact that they cannot detect the most relevant viruses.

The guideline criteria shown in Table 5 are based upon the likely viral content of source waters and the degree of treatment necessary to ensure that even very large volumes of drinking-water have a negligible risk of containing viruses.

Ground water obtained from a protected source and documented to be free from faecal contamination from its zone of influence, the well, pumps, and delivery system can be assumed to be virus-free. However, when such water is distributed, it is desirable that it is disinfected, and that a residual level of disinfectant is maintained in the distribution system to guard against contamination.

The water must meet guideline criteria for turbidity and pH (see Table 5), bacteriological quality (see Table A2.1, p. 173), and parasitological quality (see below).

## *Parasitological quality*

It is not possible to set guideline values for pathogenic protozoa, helminths, and free-living organisms, other than that these agents should not be present in drinking-water, because one or very few organisms can produce infection in humans. The analytical methods for protozoan pathogens are expensive and time-consuming and cannot be recommended for routine use. Methods for concentrating the transmission stages of *Giardia* and *Cryptosporidium* from large volumes of water are being standardized (see Volume 2). When facilities are available for studying the incidence of these parasites in surface water, these methods could be used to measure the efficiency of water treatments in removing them and the incidence of carriage of these parasites by animal vectors in the watershed. This will enable the epidemiology and zoonotic relationships of these parasites to be better understood. The control of pathogenic parasites and of other invertebrate animal life in water mains is best accomplished by proper operation and control

**Table 5. Recommended treatments for different water sources to produce water with negligible virus risk**[a]

| Type of source | Recommended treatment |
|---|---|
| **Ground water** | |
| Protected, deep wells; essentially free of faecal contamination | Disinfection[b] |
| Unprotected, shallow wells; faecally contaminated | Filtration and disinfection |
| **Surface water** | |
| Protected, impounded upland water; essentially free of faecal contamination | Disinfection |
| Unprotected impounded water or upland river; faecal contamination | Filtration and disinfection |
| Unprotected lowland rivers; faecal contamination | Pre-disinfection or storage, filtration, disinfection |
| Unprotected watershed; heavy faecal contamination | Pre-disinfection or storage, filtration, additional treatment and disinfection |
| Unprotected watershed; gross faecal contamination | Not recommended for drinking-water supply |

[a] For all sources, the median value of turbidity before terminal disinfection must not exceed 1 nephelometric turbidity unit (NTU) and must not exceed 5 NTU in single samples.

Terminal disinfection must produce a residual concentration of free chlorine of $\geq 0.5$ mg/litre after at least 30 minutes of contact in water at $pH < 8.0$, or must be shown to be an equivalent disinfection process in terms of the degree of enterovirus inactivation ($>99.99\%$).

Filtration must be either slow sand filtration or rapid filtration (sand, dual, or mixed media) preceded by adequate coagulation-flocculation (with sedimentation or flotation). Diatomaceous earth filtration or a filtration process demonstrated to be equivalent for virus reduction can also be used. The degree of virus reduction must be $>90\%$.

Additional treatment may consist of slow sand filtration, ozonation with granular activated carbon adsorption, or any other process demonstrated to achieve $>99\%$ enterovirus reduction.

[b] Disinfection should be used if monitoring has shown the presence of *E. coli* or thermotolerant coliform bacteria.

of water treatment processes and distribution practices. In particular, the attainment of the bacteriological criteria (see Table A2.1, p. 173) and the application of treatments for virological reduction (see Table 5) should, except in extraordinary cases of extreme contamination by parasites, ensure that the water has a negligible risk of transmitting parasitic diseases.

## 2.4 Monitoring

### 2.4.1 Approaches and strategies

The monitoring of drinking-water quality ideally consists of two components:

— continual control of quality on a routine basis to ascertain that treatment and distribution comply with the given objectives and regulations;
— periodic microbiological and public health surveillance of the entire water supply system from source to consumer.

The continual control function is an integral part of the responsibilities of the water supply agency, through which the waterworks management ensures the satisfactory performance of the treatment processes, the quality of the product water, and the absence of secondary contamination within the distribution network. An independent body should verify that the waterworks correctly fulfils its duties. This surveillance function usually rests with the health authorities at the local, regional, and national levels.

### 2.4.2 Sampling frequencies

The frequency of sampling will be determined by the resources available. The more frequently the water is examined, the more likely it is that chance contamination will be detected. There are two main points to be noted. Firstly, the chance of detecting pollution that occurs periodically, rather than randomly, is increased if samples are taken at different times of day and on different days of the week. Secondly, frequent examination by a simple method is more valuable than less frequent examination by a complex test or series of tests. Sampling frequencies for raw water sources will depend upon their overall quality, their size, the likelihood of contamination, and the season of the year. They should be established by local control agencies and are often specified in national regulations and guidelines. The results and information from sanitary inspection of the gathering grounds will often indicate whether increased vigilance is needed.

Sampling frequencies for treated water leaving the waterworks depend on the quality of the water source and the type of treatment. Minimum frequencies are: one sample every 2 weeks for waterworks with a ground water source; and one sample every week for waterworks with a surface water source.

The frequency of sampling must be greater where the number of people supplied is large, because of the higher number of people at risk. Advice on the design of sampling programmes and on the frequency of sampling is given in ISO standards (Table 6) and in national regulations. The minimum frequencies shown in Table 7 are recommended for water in the distribution system.

Samples should be spaced randomly within each month and from month to month, and should be taken both from fixed points, such as pumping stations and tanks, and from random locations throughout the distribution system, including points near its extremities and taps connected directly to the mains in houses and large multi-occupancy buildings, where there is a greater risk of

*Table 6. A list of International Organization for Standardization (ISO) standards for water quality giving guidance on sampling*

| ISO standard no. | Title (water quality) |
|---|---|
| 5667-1:1980 | Sampling - Part 1: Guidance on the design of sampling programmes |
| 5667-2:1982 | Sampling - Part 2: Guidance on sampling techniques |
| 5667-3:1985 | Sampling - Part 3: Guidance on the preservation and handling of samples |
| 5667-4:1987 | Sampling - Part 4: Guidance on sampling from lakes, natural and man-made |
| 5667-5:1991 | Sampling - Part 5: Guidance on sampling of drinking-water and water used for food and beverage processing |
| 5667-6:1990 | Sampling - Part 6: Guidance on sampling of rivers and streams |

*Table 7. Minimum sampling frequencies for drinking-water in the distribution system*

| Population served | Samples to be taken monthly |
|---|---|
| Less than 5000 | 1 sample |
| 5000–100 000 | 1 sample per 5000 population |
| More than 100 000 | 1 sample per 10 000 population, plus 10 additional samples |

contamination through cross-connections and back-siphonage. Frequency of sampling should be increased at times of epidemics, flooding, emergency operations, or following interruptions of supply or repair work. With systems serving small communities, periodic sanitary surveys are likely to yield more information than infrequent sampling.

No general recommendation can be made for unpiped supplies and untreated water, because the quality and likelihood of contamination will vary seasonally and with local conditions. The frequency should be established by the local control agency and reflect local conditions, including the results of sanitary surveys.

### 2.4.3 Sampling procedures

Detailed advice on the procedures to be used for sampling different sources of water or treatment plants and distribution systems and at the tap are given in

Volume 3 of *Guidelines for drinking-water quality* and in standard methods (Table 6) and other references, which should be consulted. However, the following general points should be noted.

Care must be taken to ensure that samples are representative of the water to be examined and that no accidental contamination occurs during sampling. Sample collectors should, therefore, be trained and made aware of the responsible nature of their work. Samples should be clearly labelled with the site, date, time, nature of the work, and other relevant information and sent to the laboratory for analysis without delay.

If the water to be examined is likely to contain chlorine, chloramine, chlorine dioxide, or ozone, then sodium thiosulfate solution should be added to neutralize any residual disinfectant. A properly controlled concentration of thiosulfate has no significant effect on the coliform organisms, including *E. coli*, either in chlorinated or in unchlorinated water samples during storage. If heavy metals, particularly copper, are present, then chelating agents (e.g., edetic acid (EDTA)), should also be added.

When samples of disinfected water are taken, the concentration of residual disinfectant at the sampling point and the pH should be determined at the time of collection.

When a number of samples are to be taken for various purposes from the same location, the sample for bacteriological examination should be collected first to avoid the danger of contamination of the sampling point.

Samples must be taken from different parts of the distribution system to ensure that all parts of the system are tested. When streams, lakes, or cisterns are being sampled, the water must be taken from below the surface, away from banks, sides of tanks, and stagnant zones, and without stirring up sediments. Taps, sampling ports, and the orifices of pumps should, if possible, be disinfected and a quantity of water run to waste to flush out the standing water in the pipe, before the sample is taken. Sampling ports in treatment processes and on water mains must be carefully sited, to ensure that samples are representative. The length of pipework to the tap should be as short as possible.

The changes that may occur in the bacterial content of water on storage can be reduced to a minimum by ensuring that samples are not exposed to light and are kept cool, preferably between 4 °C and 10 °C, but not frozen. Examination should begin as soon as possible after sampling and certainly within 24 hours. If samples cannot be cooled, they must be examined within 2 hours of sampling. If neither condition can be met, the sample should not be analysed. The box used to carry samples should be cleaned and disinfected after each use to avoid contaminating the surfaces of bottles and the sampler's hands.

## 2.4.4 Surveillance programme requirements

Surveillance is the continuous and vigilant public health assessment and overview of the safety and acceptability of drinking-water supplies. Each component part of the drinking-water system – the source, treatment, storage, and distribution – must function without risk of failure. A failure in one part will jeopardize and nullify the effects of other parts that function perfectly, as well as the care that has been taken to ensure that they do so. Water is liable to contamination at all stages in the process of supply, hence the need for constant vigilance. At the same time, careful and intelligent assessment of likely sources of risk and breakdown are needed before a supply is planned and installed and, indeed, continuously thereafter, because of changing conditions and potential sources of contamination. Contingency plans must be made to deal with any emergencies that may arise through natural or man-made disasters, such as accidents, hostilities and civil commotions, or cessation in supplies of essential chemicals used in treatment.

An essential part of surveillance is the establishment of a proper network for regulation and command. At the highest level, this means the establishment and enforcement of national standards, the promulgation of national guidelines for achieving compliance with the laws and standards and, at the level of the water supply agency, the promotion of local codes of good waterworks practice, together with formal instruction and training. A regulatory inspectorate, with national authority, should be established to ensure that the legal requirements are met and compliance with standards is achieved. This body should be separate from that representing the interests of the water provider.

Both the water provider and the regulatory inspectorate should have properly equipped laboratory facilities with trained and properly qualified personnel, adequate facilities for sustaining the level of monitoring required on a regular basis, and sufficient capacity to carry out additional examinations as required to meet special needs. Operational staff at the waterworks should also be appropriately trained and qualified.

Lines of communication and command must be established at the outset and must be properly understood by all staff, to the highest levels. This is to ensure effective functioning of day-to-day operations. It is also to ensure that immediate remedial action is taken when emergencies and contamination are discovered; bacteriological failures must be acted on as soon as discovered, which means that the findings of the microbiologist must carry authority with the engineer and operational staff. The lines of communication needed in an emergency will be complex, involving not only different public bodies but also geographical boundaries of responsibility. Appropriate instructions must be drawn up and understood at each site.

The scope of surveillance, with examples covering the points made in this section, has been considered in a separate WHO publication, which should be consulted (see Bibliography, p. 144). The importance of surveillance is highlighted repeatedly by official reports of serious outbreaks of waterborne disease, which usually reveal deficiencies in more than one area. Surveillance procedures are described further in Volume 3 of the *Guidelines for drinking-water quality*.

The levels of surveillance of drinking-water quality differ widely in developing countries, just as economic development and provision of community water supplies vary. Surveillance should be developed and expanded progressively, by adapting the level to the local situation and economic resources, with gradual implementation, consolidation, and development of the programme to the level ultimately desired.

# 3.
# Chemical aspects

## 3.1 Background information used

The assessment of the toxicity of drinking-water contaminants has been made on the basis of published reports from the open literature, information submitted by governments and other interested parties, and unpublished proprietary data. In the development of the guideline values, existing international approaches to developing guidelines were carefully considered. Previous risk assessments developed by the International Programme on Chemical Safety (IPCS) in Environmental Health Criteria monographs, the International Agency for Research on Cancer (IARC), the Joint FAO/WHO Meetings on Pesticide Residues (JMPR), and the Joint FAO/WHO Expert Committee on Food Additives (JECFA) were reviewed. These assessments were relied upon except where new information justified a reassessment. The quality of new data was critically evaluated prior to their use in risk assessment.

## 3.2 Drinking-water consumption and body weight

Global data on the consumption of drinking-water are limited. In studies carried out in Canada, the Netherlands, the United Kingdom, and the United States of America, the average daily *per capita* consumption was usually found to be less than 2 litres, but there was considerable variation between individuals. As water intake is likely to vary with climate, physical activity, and culture, the above studies, which were conducted in temperate zones, can give only a limited view of consumption patterns throughout the world. At temperatures above 25 °C, for example, there is a sharp rise in fluid intake, largely to meet the demands of an increased sweat rate.

In developing the guideline values for potentially hazardous chemicals, a daily *per capita* consumption of 2 litres by a person weighing 60 kg was generally assumed. The guideline values set for drinking-water using this assumption do, on average, err on the side of caution. However, such an assumption may underestimate the consumption of water per unit weight, and thus exposure, for those living in hot climates as well as for infants and children, who consume more fluid per unit weight than adults.

## 3. CHEMICAL ASPECTS

The higher intakes, and hence exposure, for infants and children apply for only a limited time, but this period may coincide with greater sensitivity to some toxic agents and less for others. Irreversible effects that occur during early age will have more social and public health significance than those that are delayed. Where it was judged that this segment of the population was at a particularly high risk from exposure to certain chemicals, the guideline value was derived on the basis of a 10-kg child consuming 1 litre per day or a 5-kg infant consuming 0.75 litre per day. The corresponding daily fluid intakes are higher than for adults on a body weight basis.

### 3.3 Inhalation and dermal absorption

The contribution of drinking-water to daily exposure includes direct ingestion as well as some indirect routes, such as inhalation of volatile substances and dermal contact during bathing or showering.

In most cases, the data were insufficient to permit reliable estimates of exposure by inhalation and dermal absorption of contaminants present in drinking-water. It was not possible, therefore, to address intake from these routes specifically in the derivation of the guideline values. However, that portion of the total tolerable daily intake (TDI) allocated to drinking-water is generally sufficient to allow for these additional routes of intake (see section 3.4.1.). When there is concern that potential inhalation of volatile compounds and dermal exposure from various indoor water uses (such as showering) are not adequately addressed, authorities could adjust the guideline value.

### 3.4 Health risk assessment

There are two principal sources of information on health effects resulting from exposure to chemicals that can be used in deriving guideline values. The first is studies on human populations. The value of such investigations is often limited, owing to lack of quantitative information on the concentrations to which people are exposed or on simultaneous exposure to other agents. The second, and the one used most often, is toxicity studies using laboratory animals. Such studies are generally limited because of the relatively small number of animals used and the relatively high doses administered. Furthermore, there is a need to extrapolate the results to the low doses to which human populations are usually exposed.

In order to derive a guideline value to protect human health, it is necessary to select the most suitable experimental animal study on which to base the extrapolation. Data from well-conducted studies, where a clear dose–response relationship has been demonstrated, are preferred. Expert judgement was exercised in the selection of the most appropriate study from the range of information available.

### 3.4.1 Derivation of guideline values using a tolerable daily intake approach

For most kinds of toxicity, it is generally believed that there is a dose below which no adverse effects will occur. For chemicals that give rise to such toxic effects, a tolerable daily intake (TDI) can be derived as follows:

$$TDI = \frac{NOAEL \text{ or } LOAEL}{UF}$$

where  $NOAEL$ = no-observed-adverse-effect level,
$LOAEL$ = lowest-observed-adverse-effect level,
$UF$ = uncertainty factor.

The guideline value (GV) is then derived from the TDI as follows:

$$GV = \frac{TDI \times bw \times P}{C}$$

where $bw$ = body weight (60 kg for adults, 10 kg for children, 5 kg for infants),
$P$ = fraction of the TDI allocated to drinking-water,
$C$ = daily drinking-water consumption (2 litres for adults, 1 litre for children, 0.75 litre for infants).

### Tolerable daily intake

The TDI is an estimate of the amount of a substance in food or drinking-water, expressed on a body weight basis (mg/kg or $\mu$g/kg of body weight), that can be ingested daily over a lifetime without appreciable health risk.

Over many years, JECFA and JMPR have developed certain principles in the derivation of acceptable daily intakes (ADIs). These principles have been adopted where appropriate in the derivation of TDIs used in developing guideline values for drinking-water quality.

ADIs are established for food additives and pesticide residues that occur in food for necessary technological purposes or plant protection reasons. For chemical contaminants, which usually have no intended function in drinking-water, the term "tolerable daily intake" is seen as more appropriate than "acceptable daily intake", as it signifies permissibility rather than acceptability.

As TDIs are regarded as representing a tolerable intake for a lifetime, they are not so precise that they cannot be exceeded for short periods of time. Short-term exposure to levels exceeding the TDI is not a cause for concern, provided the individual's intake averaged over longer periods of time does not appreciably exceed the level set. The large uncertainty factors generally involved in establishing a TDI (see below) serve to provide assurance that exposure exceeding the TDI for short periods is unlikely to have any deleterious effects upon health. However, consideration should be given to any potential acute toxic effects that may occur if the TDI is substantially exceeded for short periods of time.

The calculated TDI was used to derive the guideline value, which was then rounded to one significant figure. In some instances, ADI values with only one significant figure set by JECFA or JMPR were used to calculate the guideline value. The guideline value was generally rounded to one significant figure to reflect the uncertainty in animal toxicity data and exposure assumptions made. More than one significant figure was used for guideline values only where extensive information on toxicity and exposure to humans provided greater certainty.

### No-observed-adverse-effect level and lowest-observed-adverse-effect level

The NOAEL is defined as the highest dose or concentration of a chemical in a single study, found by experiment or observation, that causes no detectable adverse health effect. Whenever possible, the NOAEL is based on long-term studies, preferably of ingestion in drinking-water. However, NOAELs obtained from short-term studies and studies using other sources of exposure (e.g., food, air) may also be used.

If a NOAEL is not available, a LOAEL may be used, which is the lowest observed dose or concentration of a substance at which there is a detectable adverse health effect. When a LOAEL is used instead of a NOAEL, an additional uncertainty factor is normally used (see below).

### Uncertainty factors

The application of uncertainty factors has been widely used in the derivation of ADIs for food additives, pesticides, and environmental contaminants. The derivation of these factors requires expert judgement and a careful sifting of the available scientific evidence.

In the derivation of the WHO drinking-water quality guideline values, uncertainty factors were applied to the lowest NOAEL or LOAEL for the response considered to be the most biologically significant and were determined by consensus among a group of experts using the approach outlined below:

| *Source of uncertainty* | *Factor* |
|---|---|
| Interspecies variation (animals to humans) | 1–10 |
| Intraspecies variation (individual variations) | 1–10 |
| Adequacy of studies or database | 1–10 |
| Nature and severity of effect | 1–10 |

Inadequate studies or databases include those that used a LOAEL instead of a NOAEL and studies considered to be shorter in duration than desirable. Situations in which the nature or severity of effect might warrant an additional uncertainty factor include studies in which the end-point was malformation of a fetus or in which the end-point determining the NOAEL was directly related to possible carcinogenicity. In the latter case, an additional uncertainty factor was applied for carcinogenic compounds for which a guideline value was derived using a TDI approach (see section 3.4.2). Factors lower than 10 were used, for example, for interspecies variation when humans are known to be less sensitive than the animal species studied.

The total uncertainty factor should not exceed 10 000. If the risk assessment would lead to a higher uncertainty factor, then the resulting TDI would be so imprecise as to lack meaning. For substances for which uncertainty factors were greater than 1000, guideline values are designated as provisional in order to emphasize the high level of uncertainty inherent in these values.

The selection and application of uncertainty factors are important in the derivation of guideline values for chemicals, as they can make a considerable difference to the values set. For contaminants for which there is relatively little uncertainty, the guideline value was derived using a small uncertainty factor. For most contaminants, however, there is great scientific uncertainty, and a large uncertainty factor was used. Hence, there may be a large margin of safety above the guideline value before adverse health effects result.

There is considerable merit in using a method that allows a high degree of flexibility. However, it is important that, where possible, the derivation of the uncertainty factor used in calculating a guideline value is clearly presented as part of the rationale. This helps authorities in using the guidelines, as the safety margin in allowing for local circumstances is clear. It also helps in determining the urgency and nature of the action required in the event that a guideline value is exceeded.

### *Allocation of intake*

Drinking-water is not usually the sole source of human exposure to the substances for which guideline values have been set. In many cases, the intake from drinking-water is small in comparison with that from other sources such as food and air. Guideline values derived using the TDI approach take into account exposure from all sources by apportioning a percentage of the TDI to drinking-water. This approach ensures that total daily intake from all sources (including drinking-water containing concentrations of the substance at or near the guideline value) does not exceed the TDI.

Wherever possible, data concerning the proportion of total intake normally ingested in drinking-water (based on mean levels in food, air and drinking-water) or intakes estimated on the basis of consideration of physical and chemical properties were used in the derivation of the guideline values. Where such information was not available, an arbitrary (default) value of 10% for drinking-water was used. This default value is, in most cases, sufficient to account for additional routes of intake (i.e., inhalation and dermal absorption) of contaminants in water.

It is recognized that exposure from various media may vary with local circumstances. It should be emphasized, therefore, that the derived guideline values apply to a typical exposure scenario or are based on default values that may not be applicable for all areas. In those areas where relevant data on exposure are available, authorities are encouraged to develop context-specific guideline values that are tailored to local circumstances and conditions. For example, in areas where the intake of a particular contaminant in drinking-water is known to be much greater than that from other sources (i.e., air and food), it may be appropriate to allocate a greater proportion of the TDI to drinking-water to derive a guideline value more suited to the local conditions. In addition, in cases in which guideline values are exceeded, efforts should be made to assess the contribution of other sources to total intake; if practicable, exposure from these sources should be minimized.

### 3.4.2 Derivation of guideline values for potential carcinogens

The evaluation of the potential carcinogenicity of chemical substances is usually based on long-term animal studies. Sometimes data are available on carcinogenicity in humans, mostly from occupational exposure.

On the basis of the available evidence, IARC categorizes chemical substances with respect to their potential carcinogenic risk into the following groups (for a detailed description of the classifications, see box on pp. 36-37):

Group 1:  the agent is carcinogenic to humans
Group 2A: the agent is probably carcinogenic to humans
Group 2B: the agent is possibly carcinogenic to humans
Group 3:  the agent is not classifiable as to its carcinogenicity to humans
Group 4:  the agent is probably not carcinogenic to humans.

In establishing the present guideline values for drinking-water quality, the IARC classification for carcinogenic compounds was taken into consideration. For a number of compounds, additional information was also available.

It is generally considered that the initiating event in the process of chemical carcinogenesis is the induction of a mutation in the genetic material (DNA) of somatic cells (i.e., cells other than ova or sperm). Because this genotoxic mechanism theoretically does not have a threshold, there is a probability of harm at any level of exposure. Therefore, the development of a TDI is considered

### Evaluation of carcinogenic risk to humans

IARC considers the body of evidence as a whole in order to reach an overall evaluation of the carcinogenicity for humans of an agent, mixture, or circumstance of exposure.

The agent, mixture, or exposure circumstance is described according to the wording of one of the following categories, and the designated group is given. The categorization of an agent, mixture, or exposure circumstance is a matter of scientific judgement, reflecting the strength of the evidence derived from studies in humans and in experimental animals and from other relevant data.

**Group 1. The agent (mixture) is carcinogenic to humans.**
**The exposure circumstance entails exposures that are carcinogenic to humans.**

This category is used when there is *sufficient evidence* of carcinogenicity in humans. Exceptionally, an agent (mixture) may be placed in this category when evidence in humans is less than sufficient but there is *sufficient evidence* of carcinogenicity in experimental animals and strong evidence in exposed humans that the agent (mixture) acts through a relevant mechanism of carcinogenicity.

**Group 2**

This category includes agents, mixtures, and exposure circumstances for which, at one extreme, the degree of evidence of carcinogenicity in humans is almost sufficient, as well as those for which, at the other extreme, there are no human data but for which there is evidence of carcinogenicity in experimental animals. Agents, mixtures, and exposure circumstances are assigned to either group 2A (probably carcinogenic to humans) or group 2B (possibly carcinogenic to humans) on the basis of epidemiological and experimental evidence of carcinogenicity and other relevant data.

**Group 2A. The agent (mixture) is probably carcinogenic to humans.**
**The exposure circumstance entails exposures that are probably carcinogenic to humans.**

This category is used when there is *limited evidence* of carcinogenicity in humans and *sufficient evidence* of carcinogenicity in experimental animals. In some cases, an agent (mixture) may be classified in this category when there is *inadequate evidence* of carcinogenicity in humans and *sufficient evidence* of carcinogenicity in experimental animals and strong evidence that the carcinogenesis is mediated by a mechanism that also operates in humans. Exceptionally, an agent, mixture, or exposure circumstance may be classified in this category solely on the basis of *limited evidence* of carcinogenicity in humans.

## 3. CHEMICAL ASPECTS

> ***Group 2B. The agent (mixture) is possibly carcinogenic to humans. The exposure circumstance entails exposures that are possibly carcinogenic to humans.***
>
> This category is used for agents, mixtures, and exposure circumstances for which there is *limited evidence* of carcinogenicity in humans and less than *sufficient evidence* of carcinogenicity in experimental animals. It may also be used when there is *inadequate evidence* of carcinogenicity in humans but there is *sufficient evidence* of carcinogenicity in experimental animals. In some instances, an agent, mixture, or exposure circumstance for which there is *inadequate evidence* of carcinogenicity in humans but *limited evidence* of carcinogenicity in experimental animals together with supporting evidence from other relevant data may be placed in this group.
>
> ***Group 3. The agent (mixture or exposure circumstance) is not classifiable as to its carcinogenicity to humans.***
>
> This category is used most commonly for agents, mixtures, and exposure circumstances for which the evidence of carcinogenicity is inadequate in humans and inadequate or limited in experimental animals.
>
> Exceptionally, agents (mixtures) for which the evidence of carcinogenicity is inadequate in humans but sufficient in experimental animals may be placed in this category when there is strong evidence that the mechanism of carcinogenicity in experimental animals does not operate in humans.
>
> Agents, mixtures, and exposure circumstances that do not fall into any other group are also placed in this category.
>
> ***Group 4. The agent (mixture) is probably not carcinogenic to humans.***
>
> This category is used for agents or mixtures for which there is *evidence suggesting lack of carcinogenicity* in humans and in experimental animals. In some instances, agents or mixtures for which there is *inadequate evidence* of carcinogenicity in humans but *evidence suggesting lack of carcinogenicity* in experimental animals, consistently and strongly supported by a broad range of other relevant data, may be classified in this group.

inappropriate, and mathematical low-dose risk extrapolation is applied. On the other hand, there are carcinogens that are capable of producing tumours in animals or humans without exerting genotoxic activity, but acting through an indirect mechanism. It is generally believed that a threshold dose exists for these non-genotoxic carcinogens.

In order to make the distinction with respect to the underlying mechanism of carcinogenicity, each compound that has been shown to be a carcinogen was evaluated on a case-by-case basis, taking into account the evidence of genotoxicity, the range of species affected, and the relevance to humans of the tumours observed in experimental animals.

For carcinogens for which there is convincing evidence to suggest a non-genotoxic mechanism, guideline values were calculated using a TDI approach, as described in section 3.4.1.

In the case of compounds considered to be genotoxic carcinogens, guideline values were determined using a mathematical model, and the guideline values are presented as the concentration in drinking-water associated with an estimated excess lifetime cancer risk of $10^{-5}$ (one additional cancer case per 100 000 of the population ingesting drinking-water containing the substance at the guideline value for 70 years). Concentrations associated with estimated excess lifetime cancer risks of $10^{-4}$ and $10^{-6}$ can be calculated by multiplying and dividing, respectively, the guideline value by 10. In cases in which the concentration associated with a $10^{-5}$ excess lifetime cancer risk is not practical because of inadequate analytical or treatment technology, a provisional guideline value was set at a practicable level and the estimated associated cancer risk presented.

Although several models exist, the linearized multistage model was generally adopted in the development of these guidelines. As indicated in Volume 2, other models were considered more appropriate in a few cases.

It should be emphasized, however, that guideline values for carcinogenic compounds computed using mathematical models must be considered at best as a rough estimate of the cancer risk. These models do not usually take into account a number of biologically important considerations, such as pharmacokinetics, DNA repair, or immunological protection mechanisms. However, the models used are conservative and probably err on the side of caution.

To account for differences in metabolic rates between experimental animals and humans — the former are more closely correlated with the ratio of body surface areas than with body weights — a surface area to body weight correction is sometimes applied to quantitative estimates of cancer risk derived on the basis of models for low-dose extrapolation. Incorporation of this factor increases the risk by approximately one order of magnitude (depending on the species upon which the estimate is based) and increases the risk estimated on the basis of studies in mice relative to that in rats. The incorporation of this factor is considered to be overly conservative, particularly in view of the fact that linear extrapolation most likely overestimates risk at low doses; indeed, Crump et al. (1989) concluded that "all measures of dose except dose rate per unit of body weight tend to result in overestimation of human risk".[1] Consequently, guideline values for carcinogenic contaminants were developed on the basis of quantitative estimates of risk that were not corrected for the ratio of surface area to body weight.

---

[1] Crump K, Allen B, Shipp A. Choice of dose measures for extrapolating carcinogenic risk from animals to humans: an empirical investigation of 23 chemicals. *Health physics*, 1989, 57, Suppl. 1: 387-393.

## 3.5 Mixtures

Chemical contaminants of drinking-water supplies are present together with numerous other inorganic and organic constituents. The guideline values were calculated separately for individual substances, without specific consideration of the potential for interaction of each substance with other compounds present. However, the large margin of safety incorporated in the majority of guideline values is considered to be sufficient to account for such potential interactions. In addition, the majority of contaminants will not be present at concentrations at or near their guideline value.

There may, however, be occasions when a number of contaminants with similar toxicological effects are present at levels near their respective guideline values. In such cases, decisions concerning appropriate action should be made, taking into consideration local circumstances. Unless there is evidence to the contrary, it is appropriate to assume that the toxic effects of these compounds are additive.

## 3.6 Summary statements

### 3.6.1 Inorganic constituents

Aluminium
Aluminium is a widespread and abundant element, comprising some 8% of the earth's crust. Aluminium compounds are widely used as coagulants in treatment of water for public supply and the presence of aluminium in drinking-water is frequently due to deficiencies in the control and operation of the process. Human exposure may occur by a variety of routes, with drinking-water probably contributing less than 5% of the total intake.

The metabolism of aluminium in humans is not well understood, but it appears that inorganic aluminium is poorly absorbed and that most of the absorbed aluminium is rapidly excreted in the urine.

Aluminium is of low toxicity in laboratory animals, and JECFA established a provisional tolerable weekly intake (PTWI) of 7 mg/kg of body weight in 1988. However, this was based on studies of aluminium phosphate (acidic); the chemical form of aluminium in drinking-water is different.

In some studies, aluminium has appeared to be associated with the brain lesions characteristic of Alzheimer disease, and in several ecological epidemiological studies the incidence of Alzheimer disease has been associated with aluminium in drinking-water. These ecological analyses must be interpreted with caution and should be confirmed in analytical epidemiological studies.

There is a need for further studies, but the balance of epidemiological and physiological evidence at present does not support a causal role for aluminium in Alzheimer disease. Therefore, no health-based guideline value is recommended. However, a concentration of aluminium of 0.2 mg/litre in drinking-water provides a compromise between the practical use of aluminium salts in water treatment and discoloration of distributed water (see page 124).

Ammonia
The term ammonia includes the non-ionized ($NH_3$) and ionized ($NH_4^+$) species. Ammonia in the environment originates from metabolic, agricultural, and industrial processes and from disinfection with chloramine. Natural levels in ground and surface waters are usually below 0.2 mg/litre. Anaerobic ground waters may contain up to 3 mg/litre. Intensive rearing of farm animals can give rise to much higher levels in surface water. Ammonia contamination can also arise from cement mortar pipe linings. Ammonia in water is an indicator of possible bacterial, sewage, and animal waste pollution.

Ammonia is a major component of the metabolism of mammals. Exposure from environmental sources is insignificant in comparison with endogenous synthesis of ammonia. Toxicological effects are observed only at exposures above about 200 mg/kg of body weight.

Ammonia in drinking-water is not of immediate health relevance, and therefore no health-based guideline value is proposed. However, ammonia can compromise disinfection efficiency, result in nitrite formation in distribution systems, cause the failure of filters for the removal of manganese, and cause taste and odour problems (see page 124).

Antimony
Antimony salts and possibly organic complexes of antimony are typically found in food and water at low levels. Reported concentrations of antimony in drinking-water are usually less than 4 $\mu$g/litre. Estimated dietary intake for adults is about 0.02 mg/day. Where antimony–tin solder is beginning to replace lead solder, exposure to antimony may increase in the future.

In its overall evaluation based on inhalation exposure, IARC concluded that antimony trioxide is possibly carcinogenic to humans (Group 2B) and antimony trisulfide is not classifiable as to its carcinogenicity in humans (Group 3).

In a limited lifetime study in which rats were exposed to antimony in drinking-water at a single dose level of 0.43 mg/kg of body weight per day, effects observed were decreased longevity and altered blood levels of glucose and cholesterol. No effects were observed on the incidence of benign or malignant tumours.

An uncertainty factor of 500 (100 for inter- and intraspecies variation and 5 for the use of a LOAEL instead of a NOAEL) was applied to the LOAEL of 0.43 mg/kg of body weight per day, giving a TDI of 0.86 $\mu$g/kg of body weight. An allocation of 10% of the TDI to drinking-water gives a concentration of

0.003 mg/litre (rounded figure), which is below the limit of practical quantitative analysis. The provisional guideline value for antimony has therefore been set at a practical quantification level of 0.005 mg/litre. This results in a margin of safety of approximately 250-fold for potential health effects, based on the LOAEL of 0.43 mg/kg of body weight per day observed in the limited lifetime study in rats.

### Arsenic

Arsenic is widely distributed throughout the earth's crust and is used commercially, primarily in alloying agents. It is introduced into water through the dissolution of minerals and ores, from industrial effluents, and from atmospheric deposition; concentrations in ground water in some areas are sometimes elevated as a result of erosion from natural sources. The average daily intake of inorganic arsenic in water is estimated to be similar to that from food; intake from air is negligible.

Inorganic arsenic is a documented human carcinogen and has been classified by IARC in Group 1. A relatively high incidence of skin and possibly other cancers that increase with dose and age has been observed in populations ingesting water containing high concentrations of arsenic.

Arsenic has not been shown to be carcinogenic in the limited bioassays in animal species that are available, but it has given positive results in studies designed to assess the potential for tumour promotion. Arsenic has not been shown to be mutagenic in bacterial and mammalian assays, although it has been shown to induce chromosomal aberrations in a variety of cultured cell types, including human cells; such effects have not been observed *in vivo*.

Data on the association between internal cancers and ingestion of arsenic in drinking-water were insufficient for quantitative assessment of risk. Instead, owing to the documented carcinogenicity of arsenic in drinking-water in human populations, the lifetime risk of skin cancer was estimated using a multistage model. On the basis of observations in a population ingesting arsenic-contaminated drinking-water, the concentration associated with an excess lifetime skin cancer risk of $10^{-5}$ was calculated to be 0.17 $\mu$g/litre. This value may, however, overestimate the actual risk of skin cancer owing to the possible contribution of other factors to disease incidence in the population and to possible dose-dependent variations in metabolism that could not be taken into consideration. In addition, this value is below the practical quantification limit of 10 $\mu$g/litre.

With a view to reducing the concentration of this carcinogenic contaminant in drinking-water, a provisional guideline value for arsenic in drinking-water of 0.01 mg/litre is established. The estimated excess lifetime skin cancer risk associated with exposure to this concentration is $6 \times 10^{-4}$.

A similar value may be derived (assuming a 20% allocation to drinking-water) on the basis of the provisional maximum tolerable daily intake (PMTDI) for inorganic arsenic of 2 µg/kg of body weight established by JECFA in 1983 and confirmed as a PTWI of 15 µg/kg of body weight for inorganic arsenic in 1988. JECFA noted, however, that the margin between the PTWI and intakes reported to have toxic effects in epidemiological studies was narrow.

Asbestos
Asbestos is introduced into water by the dissolution of asbestos-containing minerals and ores as well as from industrial effluents, atmospheric pollution, and asbestos-cement pipes in the distribution system. Exfoliation of asbestos fibres from asbestos-cement pipes is related to the aggressiveness of the water supply. Limited data indicate that exposure to airborne asbestos released from tapwater during showers or humidification is negligible.

Asbestos is a known human carcinogen by the inhalation route. Although well studied, there has been little convincing evidence of the carcinogenicity of ingested asbestos in epidemiological studies of populations with drinking-water supplies containing high concentrations of asbestos. Moreover, in extensive studies in animal species, asbestos has not consistently increased the incidence of tumours of the gastrointestinal tract. There is, therefore, no consistent evidence that ingested asbestos is hazardous to health, and thus it was concluded that there was no need to establish a health-based guideline value for asbestos in drinking-water.

Barium
Barium occurs as a number of compounds in the earth's crust and is used in a wide variety of industrial applications, but it is present in water primarily from natural sources. In general, food is the principal source of exposure to barium; however, in areas where barium concentrations in water are high, drinking-water may contribute significantly to total intake. Intake from air is negligible.

Although an association between mortality from cardiovascular disease and the barium content of drinking-water was reported in an ecological epidemiological study, these results were not confirmed in an analytical epidemiological study of the same population. Moreover, in a short-term study in a small number of volunteers, there was no consistent indication of adverse cardiovascular effects following exposure to barium at concentrations of up to 10 mg/litre in water. There was, however, an increase in the systolic blood pressure of rats exposed to relatively low concentrations of barium in drinking-water.

A guideline value of 0.7 mg/litre (rounded figure) was derived using the NOAEL of 7.3 mg/litre from the most sensitive epidemiological study conducted to date, in which there were no significant differences in blood pressure or the prevalence of cardiovascular disease between a population drinking water containing a mean barium concentration of 7.3 mg/litre and one ingesting water

## 3. CHEMICAL ASPECTS

containing barium at 0.1 mg/litre, and incorporating an uncertainty factor of 10 to account for intraspecies variation.

This value is close to that derived on the basis of the results of toxicological studies in animal species. A TDI of 51 µg/kg of body weight was calculated, based on a NOAEL of 0.51 mg/kg of body weight per day in a chronic study in rats and incorporating uncertainty factors of 10 for intraspecies variation and 1 for interspecies variation, as the results of a well-conducted epidemiological study indicate that humans are not more sensitive than rats to barium in drinking-water. The value derived from this TDI based on 20% allocation to drinking-water would be 0.3 mg/litre (rounded figure).

The guideline value for barium in drinking-water is 0.7 mg/litre.

### Beryllium

Beryllium has a number of important minor uses, based mostly on its heat resistance. It is found infrequently in drinking-water and only at very low concentrations, usually less than 1 µg/litre.

Beryllium appears to be poorly absorbed from the gastrointestinal tract. Beryllium and beryllium compounds have been classified by IARC as being probably carcinogenic to humans (Group 2A) on the basis of occupational exposure and inhalation studies in laboratory animals. There are no adequate studies by which to judge whether it is carcinogenic by oral exposure.

Beryllium has been shown to interact with DNA and cause gene mutations, chromosomal aberrations, and sister chromatid exchange in cultured mammalian somatic cells, although it has not been shown to be mutagenic in bacterial test systems.

There are no suitable oral data on which to base a toxicologically supportable guideline value. However, the very low concentrations of beryllium normally found in drinking-water seem unlikely to pose a hazard to consumers.

### Boron

Elemental boron is used principally in composite structural materials, and boron compounds are used in some detergents and industrial processes. Boron compounds are released into water from industrial and domestic effluents. Boron is usually present in drinking-water at concentrations of below 1 mg/litre, but some higher levels have been found as a result of naturally occurring boron. The total daily intake of boron is estimated to be between 1 and 5 mg.

When administered as borate or boric acid, boron is rapidly and almost completely absorbed from the gastrointestinal tract. Boron excretion occurs mainly through the kidney.

Long-term exposure of humans to boron compounds leads to mild gastrointestinal irritation. In short- and long-term animal studies and in reproductive studies with rats, testicular atrophy has been observed. Boric acid and borates have not been shown to be mutagenic in various *in vitro* test systems. No increase in tumour incidences have been observed in long-term carcinogenicity studies in mice and rats.

A TDI of 88 µg/kg of body weight was derived by applying an uncertainty factor of 100 (for inter- and intraspecies variation) to a NOAEL, for testicular atrophy, of 8.8 mg/kg of body weight per day in a 2-year diet study in dogs. This gives a guideline value for boron of 0.3 mg/litre (rounded figure), allocating 10% of the TDI to drinking-water. It should be noted, however, that the intake of boron from food is poorly characterized and that its removal by drinking-water treatment appears to be poor.

### Cadmium

Cadmium metal is used in the steel industry and in plastics. Cadmium compounds are widely used in batteries. Cadmium is released to the environment in wastewater, and diffuse pollution is caused by contamination from fertilizers and local air pollution. Contamination in drinking-water may also be caused by impurities in the zinc of galvanized pipes and solders and some metal fittings, although levels in drinking-water are usually less than 1 µg/litre. Food is the main source of daily exposure to cadmium. The daily oral intake is 10–35 µg. Smoking is a significant additional source of cadmium exposure.

Absorption of cadmium compounds is dependent on the solubility of the compounds. Cadmium accumulates primarily in the kidneys and has a long biological half-life in humans of 10–35 years.

There is evidence that cadmium is carcinogenic by the inhalation route, and IARC has classified cadmium and cadmium compounds in Group 2A. However, there is no evidence of carcinogenicity by the oral route, and no clear evidence for the genotoxicity of cadmium.

The kidney is the main target organ for cadmium toxicity. The critical cadmium concentration in the renal cortex that would produce a 10% prevalence of low-molecular-weight proteinuria in the general population is about 200 mg/kg, and would be reached after a daily dietary intake of about 175 µg per person for 50 years.

Assuming an absorption rate for dietary cadmium of 5% and a daily excretion rate of 0.005% of body burden, JECFA concluded that, if levels of cadmium in the renal cortex are not to exceed 50 mg/kg, the total intake of cadmium should not exceed 1 µg/kg of body weight per day. The provisional tolerable weekly intake (PTWI) was therefore set at 7 µg/kg of body weight. It is recognized that the margin between the PTWI and the actual weekly intake of cadmium by the general population is small, less than 10-fold, and that this margin may be even smaller in smokers. A guideline value for cadmium of 0.003 mg/litre is established based on an allocation of 10% of the PTWI to drinking-water.

## 3. CHEMICAL ASPECTS

### Chloride

Chloride in drinking-water originates from natural sources, sewage and industrial effluents, urban run-off containing de-icing salt, and saline intrusion.

The main source of human exposure to chloride is the addition of salt to food, and the intake from this source is usually greatly in excess of that from drinking-water.

Excessive chloride concentrations increase rates of corrosion of metals in the distribution system, depending on the alkalinity of the water. This can lead to increased concentrations of metals in the supply.

No health-based guideline value is proposed for chloride in drinking-water. However, chloride concentrations in excess of about 250 mg/litre can give rise to detectable taste in water (see page 124).

### Chromium

Chromium is widely distributed in the earth's crust. It can exist in valences of +2 to +6. Total chromium concentrations in drinking-water are usually less than 2 µg/litre, although concentrations as high as 120 µg/litre have been reported. In general, food appears to be the major source of intake.

The absorption of chromium after oral exposure is relatively low and depends on the oxidation state. Chromium(VI) is more readily absorbed from the gastrointestinal tract than chromium(III) and is able to penetrate cellular membranes.

There are no adequate toxicity studies available to provide a basis for a NOAEL. In a long-term carcinogenicity study in rats given chromium(III) by the oral route, no increase in tumour incidence was observed. In rats, chromium(VI) is a carcinogen via the inhalation route, although the limited data available do not show evidence for carcinogenicity via the oral route. In epidemiological studies, an association has been found between exposure to chromium(VI) by the inhalation route and lung cancer. IARC has classified chromium(VI) in Group 1 (human carcinogen) and chromium(III) in Group 3.

Chromium(VI) compounds are active in a wide range of *in vitro* and *in vivo* genotoxicity tests, whereas chromium(III) compounds are not. The mutagenic activity of chromium(VI) can be decreased or abolished by reducing agents, such as human gastric juice.

In principle, it was considered that different guideline values for chromium(III) and chromium(VI) should be derived. However, current analytical methods favour a guideline value for total chromium.

Because of the carcinogenicity of chromium(VI) by the inhalation route and its genotoxicity, the current guideline value of 0.05 mg/litre has been questioned, but the available toxicological data do not support the derivation of a new value. As a practical measure, 0.05 mg/litre, which is considered to be unlikely to give rise to significant risks to health, has been retained as the provisional guideline value until additional information becomes available and chromium can be re-evaluated.

### Copper

Copper levels in drinking-water are usually low at only a few micrograms per litre, but copper plumbing may result in greatly increased concentrations. Concentrations can reach several milligrams per litre following a period of stagnation in pipes.

Copper is an essential element, and the intake from food is normally 1–3 mg/day. In adults, the absorption and retention rates of copper depend on the daily intake; as a consequence, copper overload is unlikely. Acute gastric irritation may be observed in some individuals at concentrations in drinking-water above 3 mg/litre. In adults with hepatolenticular degeneration, the copper regulatory mechanism is defective, and long-term ingestion can give rise to liver cirrhosis.

Copper metabolism in infants, unlike that in adults, is not well developed, and the liver of the newborn infant contains over 90% of the body burden, with much higher levels than in adults. Since 1984, there has been some concern regarding the possible involvement of copper from drinking-water in early childhood liver cirrhosis in bottle-fed infants, although this has not been confirmed.

In 1982, JECFA proposed a provisional maximum tolerable daily intake (PMTDI) of 0.5 mg/kg of body weight, based on a rather old study in dogs. With an allocation of 10% of the PMTDI to drinking-water, a provisional health-based guideline value of 2 mg/litre (rounded figure) is calculated. This study did not take into account the differences in copper metabolism in the neonate. However, a concentration of 2 mg/litre should also contain a sufficient margin of safety for bottle-fed infants, because their copper intake from other sources is usually low.

In view of the remaining uncertainties regarding copper toxicity in humans, the guideline value is considered provisional. Copper can give rise to taste problems (see page 125).

### Cyanide

The acute toxicity of cyanides is high. Cyanides can be found in some foods, particularly in some developing countries, and they are occasionally found in drinking-water, primarily as a consequence of industrial contamination.

Effects on the thyroid and particularly the nervous system were observed in some populations as a consequence of the long-term consumption of inadequately processed cassava containing high levels of cyanide. This problem seems to have decreased significantly in the West African populations in which it was widely reported, following a change in processing and a general improvement in nutritional status.

There are a very limited number of toxicological studies suitable for use in deriving a guideline value. There is, however, some indication in the literature that pigs may be more sensitive than rats. There is only one study available in

## 3. CHEMICAL ASPECTS

which a clear effect level was observed, at 1.2 mg/kg of body weight per day, in pigs exposed for 6 months. The effects observed were in behavioural patterns and serum biochemistry.

Using the LOAEL from this study and applying an uncertainty factor of 100 to reflect inter- and intraspecies variation (no additional factor for a LOAEL was considered necessary because of doubts over the biological significance of the observed changes), a TDI of 12 µg/kg of body weight was calculated.

An allocation of 20% of the TDI to drinking-water was made because exposure to cyanide from other sources is normally small and because exposure from water is only intermittent. This results in a guideline value of 0.07 mg/litre (rounded figure), which is considered to be protective for acute and long-term exposure.

### Fluoride

Fluorine accounts for about 0.3 g/kg of the earth's crust. Inorganic fluorine compounds are used in the production of aluminium, and fluoride is released during the manufacture and use of phosphate fertilizers, which contain up to 4% fluorine.

Levels of daily exposure to fluoride depend on the geographical area. If diets contain fish and tea, exposure via food may be particularly high. In specific areas, other foods and indoor air pollution may contribute considerably to total exposure. Additional intake may result from the use of fluoride toothpastes.

Exposure to fluoride from drinking-water depends greatly on natural circumstances. Levels in raw water are normally below 1.5 mg/litre, but ground water may contain about 10 mg/litre in areas rich in fluoride-containing minerals. Fluoride is sometimes added to drinking-water to prevent dental caries.

Soluble fluorides are readily absorbed in the gastrointestinal tract after intake in drinking-water.

In 1987, IARC classified inorganic fluorides in Group 3. Although there was equivocal evidence of carcinogenicity in one study in male rats, extensive epidemiological studies have shown no evidence of carcinogenicity in humans.

There is no evidence to suggest that the guideline value of 1.5 mg/litre set in 1984 needs to be revised. Concentrations above this value carry an increasing risk of dental fluorosis, and much higher concentrations lead to skeletal fluorosis. The value is higher than that recommended for artificial fluoridation of water supplies. In setting national standards for fluoride, it is particularly important to consider climatic conditions, volumes of water intake, and intake of fluoride from other sources (e.g., food, air). In areas with high natural fluoride levels, it is recognized that the guideline value may be difficult to achieve in some circumstances with the treatment technology available (see section 6.3.5).

## Hardness

Hardness in water is caused by dissolved calcium and, to a lesser extent, magnesium. It is usually expressed as the equivalent quantity of calcium carbonate.

Depending on pH and alkalinity, hardness of above about 200 mg/litre can result in scale deposition, particularly on heating. Soft waters with a hardness of less than about 100 mg/litre have a low buffering capacity and may be more corrosive to water pipes.

Although a number of ecological and analytical epidemiological studies have shown a statistically significant inverse relationship between hardness of drinking-water and cardiovascular disease, the available data are inadequate to permit a conclusion that the association is causal. There is some indication that very soft waters may have an adverse effect on mineral balance, but detailed studies were not available for evaluation.

No health-based guideline value is proposed for hardness. However, the degree of hardness in water may affect its acceptability to the consumer in terms of taste and scale deposition (see page 125).

## Hydrogen sulfide

Hydrogen sulfide is a gas with an offensive "rotten eggs" odour that is detectable at very low concentrations, below 8 $\mu g/m^3$ in air. It is formed when sulfides are hydrolysed in water. However, the level of hydrogen sulfide found in drinking-water will usually be low, because sulfides are readily oxidized in well-aerated water.

The acute toxicity to humans of hydrogen sulfide following inhalation of the gas is high; eye irritation can be observed at concentrations of 15–30 mg/m$^3$. Although oral toxicity data are lacking, it is unlikely that a person could consume a harmful dose of hydrogen sulfide from drinking-water. Consequently, no health-based guideline value is proposed. However, hydrogen sulfide should not be detectable in drinking-water by taste or odour (see page 125).

## Iron

Iron is one of the most abundant metals in the earth's crust. It is found in natural fresh waters at levels ranging from 0.5 to 50 mg/litre. Iron may also be present in drinking-water as a result of the use of iron coagulants or the corrosion of steel and cast iron pipes during water distribution.

Iron is an essential element in human nutrition. Estimates of the minimum daily requirement for iron depend on age, sex, physiological status, and iron bioavailability and range from about 10 to 50 mg/day.

As a precaution against storage in the body of excessive iron, in 1983 JECFA established a provisional maximum tolerable daily intake (PMTDI) of 0.8 mg/kg of body weight, which applies to iron from all sources except for iron oxides used as colouring agents, and iron supplements taken during pregnancy and lactation or for specific clinical requirements. An allocation of 10% of this PMTDI to drinking-water gives a value of about 2 mg/litre, which does not present a hazard to health. The taste and appearance of drinking-water will usually be affected below this level (see page 126).

No health-based guideline value for iron in drinking-water is proposed.

### Lead
Lead is used principally in the production of lead-acid batteries, solder, and alloys. The organolead compounds tetraethyl and tetramethyl lead have also been used extensively as antiknock and lubricating agents in petrol, although their use for these purposes in many countries is being phased out. Owing to the decreasing use of lead-containing additives in petrol and of lead-containing solder in the food processing industry, concentrations in air and food are declining, and intake from drinking-water constitutes a greater proportion of total intake.

Lead is present in tapwater to some extent as a result of its dissolution from natural sources, but primarily from household plumbing systems containing lead in pipes, solder, fittings, or the service connections to homes. The amount of lead dissolved from the plumbing system depends on several factors, including pH, temperature, water hardness, and standing time of the water, with soft, acidic water being the most plumbosolvent.

Placental transfer of lead occurs in humans as early as the twelfth week of gestation and continues throughout development. Young children absorb 4–5 times as much lead as adults, and the biological half-life may be considerably longer in children than in adults.

Lead is a general toxicant that accumulates in the skeleton. Infants, children up to six years of age, and pregnant women are most susceptible to its adverse health effects. Inhibition of the activity of δ-aminolaevulinic dehydratase (porphobilinogen synthase; one of the major enzymes involved in the biosynthesis of haem) in children has been observed at blood lead levels as low as 5 $\mu$g/dl, although adverse effects are not associated with its inhibition at this level. Lead also interferes with calcium metabolism, both directly and by interfering with vitamin D metabolism. These effects have been observed in children at blood lead levels ranging from 12 to 120 $\mu$g/dl, with no evidence of a threshold.

Lead is toxic to both the central and peripheral nervous systems, inducing subencephalopathic neurological and behavioural effects. There is electrophysiological evidence of effects on the nervous system in children with blood levels well below 30 $\mu$g/dl. The balance of evidence from cross-sectional epidemiological studies indicates that there are statistically significant associations between blood lead levels of 30 $\mu$g/dl and more and intelligence quotient deficits of about four points in children. Results from prospective (longitudinal) epidemiological studies suggest that prenatal exposure to lead may have early effects on mental development that do not persist to the age of 4 years. Research on primates has supported the results of the epidemiological studies, in that significant behavioural and cognitive effects have been observed following postnatal exposure resulting in blood lead levels ranging from 11 to 33 $\mu g/dl$.

Renal tumours have been induced in experimental animals exposed to high concentrations of lead compounds in the diet, and IARC has classified lead and inorganic lead compounds in Group 2B (possible human carcinogen). However, there is evidence from studies in humans that adverse neurotoxic effects other than cancer may occur at very low concentrations of lead and that a guideline value derived on this basis would also be protective for carcinogenic effects.

In 1986, JECFA established a provisional tolerable weekly intake (PTWI) for lead of 25 $\mu g/kg$ of body weight (equivalent to 3.5 $\mu g/kg$ of body weight per day) for infants and children on the basis that lead is a cumulative poison and that there should be no accumulation of body burden of lead. Assuming a 50% allocation to drinking-water for a 5-kg bottle-fed infant consuming 0.75 litres of drinking-water per day, the health-based guideline value is 0.01 mg/litre (rounded figure). As infants are considered to be the most sensitive subgroup of the population, this guideline value will also be protective for other age groups.

Lead is exceptional in that most lead in drinking-water arises from plumbing in buildings and the remedy consists principally of removing plumbing and fittings containing lead. This requires much time and money, and it is recognized that not all water will meet the guideline immediately. Meanwhile, all other practical measures to reduce total exposure to lead, including corrosion control, should be implemented.

### Manganese

Manganese is one of the more abundant metals in the earth's crust and usually occurs together with iron. Dissolved manganese concentrations in ground and surface waters that are poor in oxygen can reach several milligrams per litre. On exposure to oxygen, manganese can form insoluble oxides that may result in undesirable deposits and colour problems in distribution systems. Daily intake of manganese from food by adults is between 2 and 9 mg.

Manganese is an essential trace element with an estimated daily nutritional requirement of 30–50 $\mu g/kg$ of body weight. Its absorption rate can vary considerably according to actual intake, chemical form, and presence of other metals, such as iron and copper, in the diet. Very high absorption rates of manganese have been observed in infants and young animals.

Evidence of manganese neurotoxicity has been seen in miners following prolonged exposure to dusts containing manganese. There is no convincing evidence of toxicity in humans associated with the consumption of manganese in drinking-water, but only limited studies are available.

Intake of manganese can be as high as 20 mg/day without apparent ill effects. With an intake of 12 mg/day, a 60-kg adult would receive 0.2 mg/kg of body weight per day. Allocating 20% of the intake to drinking-water, and applying an uncertainty factor of 3 to allow for possible increased bioavailability of manganese from water, gives a value of 0.4 mg/litre.

Although no single study is suitable for use in calculating a guideline value, the weight of evidence from actual daily intake and studies in laboratory animals given manganese in drinking-water in which neurotoxic and other toxic effects were observed supports the view that a provisional health-based guideline value of 0.5 mg/litre should be adequate to protect public health.

It should be noted that manganese may be objectionable to consumers even at levels below the provisional guideline value (see page 126).

### Mercury

Mercury is present in the inorganic form in surface and ground waters at concentrations usually of less than 0.5 µg/litre. Levels in air are in the range of 2–10 ng/m$^3$. Mean dietary intake of mercury in various countries ranges from 2 to 20 µg per day per person.

The kidney is the main target organ for inorganic mercury, whereas methylmercury affects mainly the central nervous system.

In 1972, JECFA established a provisional tolerable weekly intake (PTWI) of 5 µg/kg of body weight of total mercury, of which no more than 3.3 µg/kg of body weight should be present as methylmercury. In 1988, JECFA reassessed methylmercury, as new data had become available, and confirmed the previously recommended PTWI of 3.3 µg/kg of body weight for the general population, but noted that pregnant women and nursing mothers were likely to be at greater risk from the adverse effects of methylmercury. The available data were considered insufficient to allow a specific methylmercury intake to be recommended for this population group.

To be on the conservative side, the PTWI for methylmercury was used to derive a guideline value for inorganic mercury in drinking-water. As the main exposure is from food, a 10% allocation of the PTWI to drinking-water was made. The guideline value for total mercury is 0.001 mg/litre (rounded figure).

### Molybdenum

Concentrations of molybdenum in drinking-water are usually less than 0.01 mg/litre. However, in areas near mining sites, molybdenum concentrations as high as 200 µg/litre have been reported. Dietary intake of molybdenum is about 0.1 mg per day per person. Molybdenum is considered to be an essential element, with an estimated daily requirement of 0.1–0.3 mg for adults.

No data are available on the carcinogenicity of molybdenum by the oral route.

In a 2-year study of humans exposed through their drinking-water, the NOAEL was found to be 0.2 mg/litre. There are some concerns about the quality of this study. An uncertainty factor of 10 would normally be applied to reflect intraspecies variation. However, as molybdenum is an essential element, a factor of 3 is considered to be adequate. This gives a guideline value of 0.07 mg/litre (rounded figure).

This value is within the range of that derived on the basis of results of toxicological studies in animal species and is consistent with the essential daily requirement.

### Nickel

The concentration of nickel in drinking-water is normally less than 0.02 mg/litre. Nickel released from taps and fittings may contribute up to 1 mg/litre. In special cases of release from natural or industrial nickel deposits in the ground, the nickel concentration in drinking-water may be even higher. The average daily dietary intake is normally 0.1–0.3 mg of nickel but may be as high as 0.9 mg with an intake of special food items.

The relevant database for deriving a NOAEL is limited. On the basis of a dietary study in rats in which altered organ-to-body weight ratios were observed, a NOAEL of 5 mg/kg of body weight per day was chosen. A TDI of 5 µg/kg of body weight was derived using an uncertainty factor of 1000: 100 for inter- and intraspecies variation and an extra factor of 10 to compensate for the lack of adequate studies on long-term exposure and reproductive effects, the lack of data on carcinogenicity by the oral route (although nickel, as both soluble and sparingly soluble compounds, is now considered as a human carcinogen in relation to pulmonary exposure), and a much higher intestinal absorption when taken on an empty stomach in drinking-water than when taken together with food.

With an allocation of 10% of the TDI to drinking-water, the health-based guideline value is 0.02 mg/litre (rounded figure). This value should provide sufficient protection for individuals who are sensitive to nickel.

### Nitrate and nitrite

Nitrate and nitrite are naturally occurring ions that are part of the nitrogen cycle. Naturally occurring nitrate levels in surface and ground water are generally a few milligrams per litre. In many ground waters, an increase of nitrate levels has been observed owing to the intensification of farming practice. Concentrations can reach several hundred milligrams per litre. In some countries, up to 10% of the population may be exposed to nitrate levels in drinking-water of above 50 mg/litre.

In general, vegetables will be the main source of nitrate intake when levels in drinking-water are below 10 mg/litre. When nitrate levels in drinking-water exceed 50 mg/litre, drinking-water will be the major source of total nitrate intake.

## 3. CHEMICAL ASPECTS

Experiments suggest that neither nitrate nor nitrites act directly as a carcinogen in animals, but there is some concern about increased risk of cancer in humans from the endogenous and exogenous formation of *N*-nitroso compounds, many of which are carcinogenic in animals. Suggestive evidence relating dietary nitrate exposure to cancer, especially gastric cancer, is available from geographical correlation or ecological epidemiological studies, but these results have not been confirmed in more definitive analytical studies. It must be recognized that many factors in addition to environmental nitrate exposure may be involved.

In summary, the epidemiological evidence for an association between dietary nitrate and cancer is insufficient, and the guideline value for nitrate in drinking-water is established solely to prevent methaemoglobinaemia, which depends upon the conversion of nitrate to nitrite. Although bottle-fed infants of less than 3 months of age are most susceptible, occasional cases have been reported in some adult populations.

Extensive epidemiological data support the current guideline value for nitrate-nitrogen of 10 mg/litre. However, this value should not be expressed on the basis of nitrate-nitrogen but on the basis of nitrate itself, which is the chemical entity of concern to health, and the guideline value for nitrate is therefore 50 mg/litre.

As a result of recent evidence of the presence of nitrite in some water supplies, it was concluded that a guideline value for nitrite should be proposed. However, the available animal studies are not appropriate for the establishment of a firm NOAEL for methaemoglobinaemia in rats. Therefore, a pragmatic approach was followed, accepting a relative potency for nitrite and nitrate with respect to methaemoglobin formation of 10:1 (on a molar basis). On this basis, a provisional guideline value for nitrite of 3 mg/litre is proposed. Because of the possibility of simultaneous occurrence of nitrite and nitrate in drinking-water, the sum of the ratios of the concentration of each to its guideline value should not exceed 1, i.e.

$$\frac{C_{nitrite}}{GV_{nitrite}} + \frac{C_{nitrate}}{GV_{nitrate}} \leq 1$$

where $C$ = concentration
$GV$ = guideline value.

Dissolved oxygen
No health-based guideline value is recommended for dissolved oxygen in drinking-water. However, a dissolved oxygen content substantially lower than the saturation concentration may be indicative of poor water quality (see page 126).

pH
No health-based guideline value is proposed for pH, although eye irritation and

exacerbation of skin disorders have been associated with pH values greater than 11. Although pH usually has no direct impact on consumers, it is one of the most important operational water quality parameters (see page 127).

Selenium

Selenium levels in drinking-water vary greatly in different geographical areas but are usually much less than 0.01 mg/litre. Foodstuffs such as cereals, meat, and fish are the principal source of selenium in the general population. Levels in food vary greatly according to geographical area of production.

Selenium is an essential element for humans and forms an integral part of the enzyme glutathione peroxidase and probably other proteins as well. Most selenium compounds are water-soluble and are efficiently absorbed from the intestine. The toxicity of selenium compounds appears to be of the same order in both humans and laboratory animals.

Except for selenium sulfide, which does not occur in drinking-water, experimental data do not indicate that selenium is carcinogenic. IARC has placed selenium and selenium compounds in Group 3. Selenium compounds have been shown to be genotoxic in *in vitro* systems with metabolic activation, but not in humans. This effect may be dose-dependent *in vivo*. There is no evidence of teratogenic effects in monkeys, but no data exist for humans.

Long-term toxicity in rats is characterized by depression of growth and liver pathology at selenium levels of 0.03 mg/kg of body weight per day given in food.

In humans, the toxic effects of long-term selenium exposure are manifested in nails, hair and liver. Data from China indicate that clinical signs occur at a daily intake above 0.8 mg. Daily intakes of Venezuelan children with clinical signs were estimated to be about 0.7 mg, on the basis of their blood levels and the Chinese data on the relationship between blood level and intake. Effects on synthesis of a liver protein were also seen in a small group of patients with rheumatoid arthritis given selenium at a rate of 0.25 mg/day in addition to selenium from food. No clinical or biochemical signs of selenium toxicity were reported in a group of 142 persons with a mean daily intake of 0.24 mg (maximum 0.72 mg).

On the basis of these data, the NOAEL in humans was estimated to be about 4 $\mu$g/kg of body weight per day. The recommended daily intake of selenium is about 1 $\mu$g/kg of body weight for adults. An allocation of 10% of the NOAEL in humans to drinking-water gives a health-based guideline value of 0.01 mg/litre (rounded figure).

Silver

Silver occurs naturally mainly in the form of its very insoluble and immobile oxides, sulfides, and some salts. It has occasionally been found in ground, surface, and drinking-water at concentrations above 5 µg/litre. Levels in drinking-water treated with silver for disinfection (see section 6.3.4) may be above 50 µg/litre. Recent estimates of daily intake are about 7 µg per person.

Only a small percentage of silver is absorbed. Retention rates in humans and laboratory animals range between 0 and 10%.

The only obvious sign of silver overload is argyria, a condition in which skin and hair are heavily discoloured by silver in the tissues. An oral NOAEL for argyria in humans for a total lifetime intake of 10 g of silver was estimated on the basis of human case reports and long-term animal experiments.

The low levels of silver in drinking-water, generally below 5 µg/litre, are not relevant to human health with respect to argyria. On the other hand, special situations exist where silver salts may be used to maintain the bacteriological quality of drinking-water. Higher levels of silver, up to 0.1 mg/litre (this concentration gives a total dose over 70 years of half the human NOAEL of 10 g), could be tolerated in such cases without risk to health.

No health-based guideline value is proposed for silver in drinking-water.

### Sodium

Sodium salts (e.g., sodium chloride) are found in virtually all food (the main source of daily exposure) and drinking-water. Although concentrations of sodium in potable water are typically less than 20 mg/litre, they can greatly exceed this in some countries. The levels of sodium salts in air are normally low in relation to those in food or water. It should be noted that some water softeners can add significantly to the sodium content of drinking-water.

No firm conclusions can be drawn concerning the possible association between sodium in drinking-water and the occurrence of hypertension. Therefore, no health-based guideline value is proposed. However, concentrations in excess of 200 mg/litre may give rise to unacceptable taste (see page 127).

### Sulfate

Sulfates occur naturally in numerous minerals and are used commercially, principally in the chemical industry. They are discharged into water in industrial wastes and through atmospheric deposition; however, the highest levels usually occur in ground water and are from natural sources. In general, food is the principal source of exposure to sulfate, although intake from drinking-water can exceed that from food in areas with high concentrations. The contribution of air to total intake is negligible.

Sulfate is one of the least toxic anions; however, catharsis, dehydration, and gastrointestinal irritation have been observed at high concentrations. Magnesium sulfate, or Epsom salts, has been used as a cathartic for many years.

No health-based guideline is proposed for sulfate. However, because of the gastrointestinal effects resulting from ingestion of drinking-water containing high sulfate levels, it is recommended that health authorities be notified of sources of drinking-water that contain sulfate concentrations in excess of 500 mg/litre. The presence of sulfate in drinking-water may also cause noticeable taste (see page 127) and may contribute to the corrosion of distribution systems.

### Inorganic tin

Tin is used principally in the production of coatings used in the food industry. Food, particularly canned food, therefore represents the major route of human exposure to tin. For the general population, drinking-water is not a significant source of tin, and levels in drinking-water greater than 1–2 µg/litre are exceptional. However, there is increasing use of tin in solder, which may be used in domestic plumbing.

Tin and inorganic tin compounds are poorly absorbed from the gastrointestinal tract, do not accumulate in tissues, and are rapidly excreted, primarily in the faeces.

No increased incidence of tumours was observed in long-term carcinogenicity studies conducted in mice and rats fed stannous chloride. Tin has not been shown to be teratogenic or fetotoxic in mice, rats, and hamsters. In rats, the NOAEL in a long-term feeding study was 20 mg/kg of body weight per day.

The main adverse effect on humans of excessive levels of tin in foods (above 150 mg/kg), such as canned fruit, has been acute gastric irritation. There is no evidence of adverse effects in humans associated with chronic exposure to tin.

It was concluded that, because of the low toxicity of inorganic tin, a tentative guideline value could be derived three orders of magnitude higher than the normal tin concentration in drinking-water. Therefore, the presence of tin in drinking-water does not represent a hazard to human health. For this reason, the establishment of a numerical guideline value for inorganic tin is not deemed necessary.

### Total dissolved solids

Total dissolved solids (TDS) comprise inorganic salts (principally calcium, magnesium, potassium, sodium, bicarbonates, chlorides and sulfates) and small amounts of organic matter that are dissolved in water. TDS in drinking-water originate from natural sources, sewage, urban run-off, and industrial wastewater. Salts used for road de-icing in some countries may also contribute to the TDS content of drinking-water. Concentrations of TDS in water vary considerably in different geological regions owing to differences in the solubilities of minerals.

Reliable data on possible health effects associated with the ingestion of TDS in drinking-water are not available, and no health-based guideline value is proposed. However, the presence of high levels of TDS in drinking-water may be objectionable to consumers (see page 127).

3. CHEMICAL ASPECTS

Uranium
Uranium is present in the earth's crust, principally in the hexavalent form. It is used primarily as a fuel in nuclear energy plants and is introduced into water supplies as a result of leaching from natural sources, from mill tailings, from emissions from the nuclear industry, from the combustion of coal and other fuels, and from phosphate fertilizers. Although available information on concentrations in food and drinking-water is limited, it is likely that food is the principal source of intake of uranium in most areas.

Uranium accumulates in the kidney, and nephropathy is the primary induced effect in humans and animals. In experimental animals, uranium most commonly causes damage to the proximal convoluted tubules of the kidney, predominantly in the distal two-thirds. At doses that are not high enough to destroy a critical mass of kidney cells, the effect is reversible, as some of the lost cells are replaced.

Adequate short- and long-term studies on the chemical toxicity of uranium are not available, and therefore a guideline value for uranium in drinking-water was not derived. Until such information becomes available, it is recommended that the limits for radiological characteristics of uranium be used (see Chapter 4). The equivalent for natural uranium, based on these limits, is approximately 140 $\mu$g/litre.

Zinc
Zinc is an essential trace element found in virtually all food and potable water in the form of salts or organic complexes. The diet is normally the principal source of zinc. Although levels of zinc in surface and ground water normally do not exceed 0.01 and 0.05 mg/litre, respectively, concentrations in tapwater can be much higher as a result of dissolution of zinc from pipes.

In 1982, JECFA proposed a provisional maximum tolerable daily intake for zinc of 1 mg/kg of body weight. The daily requirement for adult men is 15–20 mg/day. It was concluded that, taking into account recent studies on humans, the derivation of a health-based guideline value is not required at this time. However, drinking-water containing zinc at levels above 3 mg/litre may not be acceptable to consumers (see page 128).

### 3.6.2 Organic constituents

**Chlorinated alkanes**

Carbon tetrachloride
Carbon tetrachloride is used principally in the production of chlorofluorocarbon refrigerants. It is released into air and water during manufacturing and use. Although available data on concentrations in food are limited, the intake of carbon tetrachloride from air is expected to be much greater than that from food

or drinking-water. Concentrations in drinking-water are generally less than 5 µg/litre.

Carbon tetrachloride has been classified in Group 2B by IARC. It can be metabolized in microsomal systems to a trichloromethyl radical that binds to macromolecules, initiating lipid peroxidation and destroying cell membranes. It has been shown to cause hepatic and other tumours in rats, mice, and hamsters after oral, subcutaneous, and inhalation exposure. The time to first tumour has sometimes been short, within 12–16 weeks in some experiments.

Carbon tetrachloride has not been shown to be mutagenic in bacterial tests with or without metabolic activation, nor has it been shown to induce effects on chromosomes or unscheduled DNA synthesis in mammalian cells either *in vivo* or *in vitro*. It has induced point mutations and gene recombination in a eukaryotic test system.

Carbon tetrachloride, therefore, has not been shown to be genotoxic in most available studies, and it is possible that it acts as a non-genotoxic carcinogen. The NOAEL in a 12-week oral gavage study in rats was 1 mg/kg of body weight per day. A TDI of 0.714 µg/kg of body weight (allowing for 5 days per week dosing) was calculated by applying an uncertainty factor of 1000 (100 for intra- and interspecies variation, and 10 for evidence of possibly non-genotoxic carcinogenicity). No additional factor for the short duration of the study was incorporated. It was considered to be unnecessary because the compound was administered in corn oil in the critical study and available data indicate that the toxicity following administration in water may be an order of magnitude less. The guideline value derived from this TDI, based on 10% allocation to drinking-water, is 2 µg/litre (rounded figure).

### Dichloromethane

Dichloromethane, or methylene chloride, is widely used as a solvent for many purposes, including coffee decaffeination and paint stripping. Exposure from drinking-water is likely to be insignificant compared with other sources.

Dichloromethane is of low acute toxicity. An inhalation study in mice provided conclusive evidence of carcinogenicity, whereas a drinking-water study provided only suggestive evidence. IARC has placed dichloromethane in Group 2B; however, the balance of evidence suggests that it is not a genotoxic carcinogen and that genotoxic metabolites are not formed in relevant amounts *in vivo*.

A TDI of 6 µg/kg of body weight was calculated by applying an uncertainty factor of 1000 (100 for inter- and intraspecies variation and 10 reflecting concern about carcinogenic potential) to a NOAEL of 6 mg/kg of body weight per day for hepatotoxic effects in a 2-year drinking-water study in rats. This gives a guideline value of 20 µg/litre (rounded figure), allocating 10% of the TDI to drinking-water. It should be noted that widespread exposure from other sources is possible.

### 1,1-Dichloroethane

1,1-Dichloroethane is used as a chemical intermediate and solvent. There are limited data showing that it can be present in concentrations of up to 10 µg/litre in drinking-water. However, because of the widespread use and disposal of this chemical, its occurrence in ground water may increase.

1,1-Dichloroethane is rapidly metabolized by mammals to acetic acid and a variety of chlorinated compounds. It is of relatively low acute toxicity, and limited data are available on its toxicity from short- and long-term studies.

There is limited *in vitro* evidence of genotoxicity. One carcinogenicity study by gavage in mice and rats provided no conclusive evidence of carcinogenicity, although there was some evidence of an increased incidence of haemangiosarcomas in treated animals.

In view of the very limited database on toxicity and carcinogenicity, it was concluded that no guideline value should be proposed.

### 1,2-Dichloroethane

1,2-Dichloroethane is used mainly as an intermediate in the production of vinyl chloride and other chemicals and to a lesser extent as a solvent. It has been found in drinking-water at levels of up to a few micrograms per litre. It is found in urban air.

IARC has classified 1,2-dichloroethane in Group 2B. It has been shown to produce statistically significant increases in a number of tumour types in laboratory animals, including the relatively rare haemangiosarcoma, and the balance of evidence indicates that it is potentially genotoxic. There are no suitable long-term studies on which to base a TDI.

On the basis of haemangiosarcomas observed in male rats in a 78-week gavage study, and applying the linearized multistage model, a guideline value for drinking-water of 30 µg/litre, corresponding to an excess lifetime cancer risk of $10^{-5}$, was calculated.

### 1,1,1-Trichloroethane

1,1,1-Trichloroethane has been found in only a small proportion of surface and ground waters, usually at concentrations of less than 20 µg/litre. In a few instances, much higher concentrations have been observed. There appears to be increasing exposure to 1,1,1-trichloroethane.

1,1,1-Trichloroethane is rapidly absorbed from the lungs and gastrointestinal tract, but only small amounts – about 6% in humans and 3% in experimental animals – are metabolized. Exposure to high concentrations can lead to hepatic steatosis (fatty liver) in both humans and laboratory animals.

IARC has placed 1,1,1-trichloroethane in Group 3. Available studies of oral administration were considered inadequate for calculation of a TDI. As there is an increasing need for guidance on this compound, a 14-week inhalation study in male mice was selected for use in calculating the guideline value. Based on a NOAEL of 1365 mg/m$^3$, a TDI of 580 µg/kg of body weight was calculated from a total absorbed dose of 580 mg/kg of body weight per day (assuming an average mouse body weight of 30 g, breathing rate of 0.043 m$^3$/day, and absorption of 30% of the air concentration), applying an uncertainty factor of 1000 (100 for inter- and intraspecies variation and 10 for the short duration of the study). A provisional guideline value of 2000 µg/litre (rounded value) is proposed, allocating 10% of the TDI to drinking-water.

This value is provisional because of the use of an inhalation study rather than an oral study. It is strongly recommended that an adequate oral toxicity study be conducted to provide more acceptable data for the derivation of a guideline value.

### Chlorinated ethenes

#### Vinyl chloride

Vinyl chloride is used primarily for the production of polyvinyl chloride. The background level of vinyl chloride in ambient air in western Europe is estimated to range from 0.1 to 0.5 µg/m$^3$. Residual vinyl chloride levels in food and drinks are now below 10 µg/kg. Vinyl chloride has been found in drinking-water at levels of up to a few micrograms per litre, and, on occasion, much higher concentrations have been found in ground water. It can be formed in water from trichloroethene and tetrachloroethene.

Vinyl chloride is metabolized to highly reactive and mutagenic metabolites by a dose-dependent and saturable pathway.

The acute toxicity of vinyl chloride is low, but vinyl chloride is toxic to the liver after short- and long-term exposure to low concentrations. Vinyl chloride has been shown to be mutagenic in various test systems *in vitro* and *in vivo*.

There is sufficient evidence of the carcinogenicity of vinyl chloride in humans from industrial populations exposed to high concentrations, and IARC has classified vinyl chloride in Group 1. A causal association between vinyl chloride exposure and angiosarcoma of the liver is sufficiently proved. Some studies suggest that vinyl chloride is also associated with hepatocellular carcinoma, brain tumours, lung tumours, and malignancies of the lymphatic and haematopoietic tissues.

Animal data show vinyl chloride to be a multisite carcinogen. Vinyl chloride administered orally or by inhalation to mice, rats, and hamsters produced tumours in the mammary gland, lungs, Zymbal gland, and skin, as well as angiosarcomas of the liver and other sites.

Because there are no data on carcinogenic risk following oral exposure of humans to vinyl chloride, estimation of risk of cancer in humans was based on animal carcinogenicity bioassays involving oral exposure. Using results from the rat bioassay, which yields the most protective value, and applying the linearized multistage model, the human lifetime exposure for a $10^{-5}$ excess risk of hepatic angiosarcoma was calculated to be 20 µg per person per day. It was also assumed that, in humans, the number of cancers at other sites may equal that of angiosarcoma of the liver, so that a correction (factor of 2) for cancers other than angiosarcoma is justified. Using the lifetime exposure of 20 µg per person per day for a $10^{-5}$ excess risk of hepatic angiosarcoma, a guideline value for drinking-water of 5 µg/litre was calculated.

### 1,1-Dichloroethene

1,1-Dichloroethene, or vinylidene chloride, is an occasional contaminant of drinking-water. It is usually found together with other chlorinated hydrocarbons. There are no data on levels in food, but levels in air are generally less than 40 ng/m³ except at some manufacturing sites.

Following oral or inhalation exposure, 1,1-dichloroethene is almost completely absorbed, extensively metabolized, and rapidly excreted. It is a central nervous system depressant and may cause liver and kidney toxicity in occupationally exposed humans. It causes liver and kidney damage in laboratory animals.

IARC has placed 1,1-dichloroethene in Group 3. It was found to be genotoxic in a number of test systems *in vitro* but was not active in the dominant lethal assay *in vivo*. It induced kidney tumours in mice in one inhalation study but was reported not to be carcinogenic in a number of other studies, including several in which it was given in drinking-water.

A TDI of 9 µg/kg of body weight was calculated from a LOAEL of 9 mg/kg of body weight per day in a 2-year drinking-water study in rats, using an uncertainty factor of 1000 (100 for intra- and interspecies variation and 10 for the use of a LOAEL in place of a NOAEL and the potential for carcinogenicity). This gives a guideline value of 30 µg/litre (rounded figure) for a 10% contribution to the TDI from drinking-water.

### 1,2-Dichloroethene

1,2-Dichloroethene exists in a *cis* and a *trans* form. The *cis* form is more frequently found as a water contaminant. The presence of these two isomers, which are metabolites of other unsaturated halogenated hydrocarbons in wastewater and anaerobic ground water, may indicate the simultaneous presence of more toxic organochlorine chemicals, such as vinyl chloride. Accordingly, their presence indicates that more intensive monitoring should be conducted. There are no data on exposure from food. Concentrations in air are low, with higher concentrations, in the microgram per cubic metre range, near production sites. The *cis*-isomer was previously used as an anaesthetic.

There is little information on the absorption, distribution, and excretion of 1,2-dichloroethene. However, by analogy with 1,1-dichloroethene, it would be expected to be readily absorbed, distributed mainly to the liver, kidneys, and lungs, and rapidly excreted. The *cis*-isomer is more rapidly metabolized than the *trans*-isomer in *in vitro* systems.

Both isomers have been reported to cause increased serum alkaline phosphatase levels in rodents. In a 3-month study in mice given the *trans*-isomer in drinking-water, there was a reported increase in serum alkaline phosphatase and reduced thymus and lung weights. Transient immunological effects were also reported, the toxicological significance of which is unclear. *Trans*-1,2-dichloroethene also caused reduced kidney weights in rats, but at higher doses. Only one rat toxicity study is available for the *cis*-isomer, which produced toxic effects in rats similar in magnitude to those induced by the *trans*-isomer in mice, but at higher doses.

There are limited data to suggest that both isomers may possess some genotoxic activity. There is no information on carcinogenicity.

Data on the *trans*-isomer were used to calculate a joint guideline value for both isomers because toxicity for the *trans*-isomer occurred at a lower dose than for the *cis*-isomer and because data suggest that the mouse is a more sensitive species than the rat. Accordingly, the NOAEL of 17 mg/kg of body weight per day from the *trans*-isomer toxicity study in mice was used to calculate a guideline value. An uncertainty factor of 1000 (100 for intra- and interspecies variation and 10 for the short duration of the study) was applied to derive a TDI of 17 µg/kg of body weight, giving a guideline value of 50 µg/litre (rounded figure) for an allocation of 10% of the TDI to drinking-water.

Trichloroethene
Trichloroethene is used mainly in dry cleaning and in metal-degreasing operations. Its use in industrialized countries has declined sharply since 1970. It is released mainly to the atmosphere but may be introduced into surface and ground water in industrial effluents. It is expected that exposure to trichloroethene from air will be greater than that from food or drinking-water. Trichloroethene in anaerobic ground water may degrade to more toxic compounds, including vinyl chloride.

Trichloroethene is rapidly absorbed from the lungs and gastrointestinal tract and distributed to all tissues. Humans metabolize between 40% and 75% of retained trichloroethene. Urinary metabolites include trichloroacetaldehyde, trichloroethanol, and trichloroacetic acid; the reactive epoxide trichloroethene oxide is an essential feature of the metabolic pathway.

Trichloroethene has been classified by IARC in Group 3. It has been shown to induce lung and liver tumours in various strains of mice at toxic doses. However, there are no conclusive data that this chemical causes cancer in other species. Trichloroethene is a weakly active mutagen in bacteria and yeast.

A TDI of 23.8 µg/kg of body weight (including allowance for 5 days per week dosing) was therefore calculated by applying an uncertainty factor of 3000 to a LOAEL of 100 mg/kg of body weight per day for minor effects on relative liver weight in a 6-week study in mice. The uncertainty factor components are 100 for inter- and intraspecies variation, 10 for limited evidence of carcinogenicity, and an additional factor of 3 in view of the short duration of the particular study and the use of a LOAEL rather than a NOAEL. The provisional guideline value derived from this TDI, based on 10% allocation to drinking-water, is 70 µg/litre (rounded figure).

### Tetrachloroethene

Tetrachloroethene has been used primarily as a solvent in dry-cleaning industries and to a lesser extent as a degreasing solvent. Tetrachloroethene is widespread in the environment and is found in trace amounts in water, aquatic organisms, air, foodstuffs, and human tissue. The highest environmental levels of tetrachloroethene are found in the commercial dry-cleaning and metal-degreasing industries. Emissions can sometimes lead to high concentrations in ground water. Tetrachloroethene in anaerobic ground water may degrade to more toxic compounds, including vinyl chloride.

At high concentrations, tetrachloroethene causes central nervous system depression. Lower concentrations of tetrachloroethene have been reported to damage the liver and the kidneys.

IARC has classified tetrachloroethene in Group 2B. It has been reported to produce liver tumours in male and female mice, with some evidence of mononuclear cell leukaemia in male and female rats and kidney tumours in male rats. The overall evidence from studies conducted to assess genotoxicity of tetrachloroethene, including induction of single-strand DNA breaks, mutation in germ cells, and chromosomal aberrations *in vitro* and *in vivo*, indicates that tetrachloroethene is not genotoxic.

In view of the overall evidence for non-genotoxicity and evidence for a saturable metabolic pathway leading to kidney tumours in rats, it is appropriate to use a NOAEL with a suitable uncertainty factor. A 6-week gavage study in male mice and a 90-day drinking-water study in male and female rats both indicated a NOAEL for hepatotoxic effects of 14 mg/kg of body weight per day. A TDI of 14 µg/kg of body weight was calculated by applying an uncertainty factor of 1000 (100 for intra- and interspecies variation and an additional 10 for carcinogenic potential). In view of the database on tetrachloroethene and considerations regarding the application of the dose via drinking-water in one of the two critical studies, it was considered unnecessary to include an additional uncertainty factor to reflect the length of the study. The guideline value for tetrachloroethene is 40 µg/litre (rounded figure) for a drinking-water contribution of 10%.

### Aromatic hydrocarbons

Benzene
Benzene is used principally in the production of other organic chemicals. It is present in petrol, and vehicular emissions constitute the main source of benzene in the environment. Benzene may be introduced into water by industrial effluents and atmospheric pollution. Concentrations in drinking-water are generally less than 5 µg/litre.

Acute exposure of humans to high concentrations of benzene primarily affects the central nervous system. At lower concentrations, benzene is toxic to the haematopoietic system, causing a continuum of haematological changes, including leukaemia. Because it is carcinogenic to humans, IARC has classified benzene in Group 1.

Haematological abnormalities similar to those observed in humans have been observed in animal species exposed to benzene. In animal studies, benzene was shown to be carcinogenic following both inhalation and ingestion. It induced several types of tumours in both rats and mice in a 2-year carcinogenesis bioassay by gavage in corn oil. Benzene has not been found to be mutagenic in bacterial assays but has been shown to cause chromosomal aberrations *in vivo* in a number of species, including humans, and to be positive in the mouse micronucleus test.

Because of the unequivocal evidence of the carcinogenicity of benzene in humans and laboratory animals and its documented chromosomal effects, quantitative risk extrapolation was used to calculate lifetime cancer risks. Based on a risk estimate using data on leukaemia from epidemiological studies involving inhalation exposure, it was calculated that a drinking-water concentration of 10 µg/litre was associated with an excess lifetime cancer risk of $10^{-5}$.

As data on the carcinogenic risk to humans following ingestion of benzene are not available, risk estimates were also carried out on the basis of the 2-year gavage study in rats and mice. The robust linear extrapolation model was used because there was a statistical lack of fit of some of the data with the linearized multistage model. The estimated range of concentrations in drinking-water corresponding to an excess lifetime cancer risk of $10^{-5}$, based on leukaemia and lymphomas in female mice and oral cavity squamous cell carcinomas in male rats, is 10–80 µg/litre. The lower end of this estimate corresponds to the estimate derived from epidemiological data, which formed the basis for the previous guideline value of 10 µg/litre associated with a $10^{-5}$ excess lifetime cancer risk. This guideline value of 10 µg/litre, for a $10^{-5}$ excess cancer risk, is therefore retained.

### Toluene

Toluene is used primarily as a solvent and in blending petrol. Concentrations of a few micrograms per litre have been found in surface water, ground water, and drinking-water. Point emissions can lead to higher concentrations in ground water. The main exposure is via air. Exposure is increased by smoking and in traffic.

Toluene is absorbed completely from the gastrointestinal tract and rapidly distributed in the body with a preference for adipose tissue. Toluene is rapidly metabolized and, following conjugation, excreted predominantly in urine.

With occupational exposure, impairment of the central nervous system and irritation of mucous membranes are observed. The acute oral toxicity is low. Toluene exerts embryotoxic and fetotoxic effects, but there is no clear evidence for teratogenic activity in laboratory animals and humans.

In long-term inhalation studies in rats and mice, there was no evidence for carcinogenicity of toluene. Genotoxicity tests *in vitro* were negative, whereas *in vivo* assays showed conflicting results with respect to chromosomal aberrations.

A TDI of 223 µg/kg of body weight was derived using a LOAEL for marginal hepatotoxic effects of 312 mg/kg of body weight per day in a 13-week gavage study in mice (administration 5 days per week) and applying an uncertainty factor of 1000 (100 for inter- and intraspecies variation and 10 for the short duration of the study and use of a LOAEL instead of a NOAEL). This yields a guideline value of 700 µg/litre (rounded figure), allocating 10% of the TDI to drinking-water. It should be noted, however, that this value exceeds the lowest reported odour threshold for toluene in water (see page 128).

### Xylenes

Xylenes are used in blending petrol, as a solvent, and as a chemical intermediate. They are released to the environment largely via air.

Concentrations of up to 8 µg/litre have been reported in surface water, ground water, and drinking-water. Levels of a few milligrams per litre were found in ground water polluted by point emissions. Exposure to xylenes is mainly from air, and exposure is increased by smoking.

Xylenes are rapidly absorbed by inhalation. Data on oral exposure are lacking. Xylenes are rapidly distributed in the body, predominantly in adipose tissue. They are almost completely metabolized and excreted in urine.

The acute oral toxicity of xylenes is low. No convincing evidence for teratogenicity has been found. Long-term carcinogenicity studies have shown no evidence for carcinogenicity. *In vitro* as well as *in vivo* mutagenicity tests have proved negative.

A TDI of 179 µg/kg of body weight was derived using a NOAEL of 250 mg/kg of body weight per day based on decreased body weight in a 103-week gavage study in rats (administration 5 days per week), applying an uncertainty factor of 1000 (100 for intra- and interspecies variation and 10 for the limited toxicological end-point). This yields a guideline value of 500 µg/litre (rounded figure), allocating 10% of the TDI to drinking-water. This value exceeds the lowest reported odour threshold for xylenes in drinking-water (see page 128).

Ethylbenzene
The primary sources of ethylbenzene in the environment are the petroleum industry and the use of petroleum products.

Because of its physical and chemical properties, more than 96% of ethylbenzene in the environment can be expected to be present in air. Values of up to 26 µg/m$^3$ in air have been reported. It is found in trace amounts in surface water, ground water, drinking-water, and food.

Ethylbenzene is readily absorbed by oral, inhalation, or dermal routes. In humans, storage in fat has been reported. Ethylbenzene is almost completely converted to soluble metabolites, which are excreted rapidly in urine.

The acute oral toxicity is low. No definite conclusions can be drawn from limited teratogenicity data. No data on reproduction, long-term toxicity, or carcinogenicity are available. Ethylbenzene has shown no evidence of genotoxicity in *in vitro* or *in vivo* systems.

A TDI of 97.1 µg/kg of body weight was derived using a NOAEL of 136 mg/kg of body weight per day, corrected for 5 days per week dosing, based on hepatotoxicity and nephrotoxicity observed in a limited 6-month study in rats, and applying an uncertainty factor of 1000 (100 for inter- and intraspecies variation and 10 for the limited database and short duration of the study). This yields a guideline value of 300 µg/litre (rounded figure), allocating 10% of the TDI to drinking-water. This value exceeds the lowest reported odour threshold for ethylbenzene in drinking-water (see page 128).

Styrene
Styrene, which is used primarily for the production of plastics and resins, is found in trace amounts in surface water, drinking-water, and food. In industrial areas, exposure levels from air can be a few hundred micrograms per day. Smoking may increase daily exposure by up to 10-fold.

Following oral or inhalation exposure, styrene is rapidly absorbed and widely distributed in the body, with a preference for lipid depots. It is metabolized to the active intermediate styrene-7,8-oxide, which is conjugated with glutathione or further metabolized. Metabolites are rapidly and almost completely excreted in urine.

Styrene has low acute toxicity. With occupational exposure, irritation of mucous membranes, depression of the central nervous system, and possibly hepatotoxicity can occur. In short-term toxicity studies in rats, impairment of glutathione transferase activity and reduced glutathione concentrations were observed.

In *in vitro* tests, styrene has been shown to be mutagenic in the presence of metabolic activation only. In *in vitro* as well as in *in vivo* studies, chromosomal aberrations have been observed, mostly at high doses of styrene. The reactive intermediate styrene-7,8-oxide is a direct-acting mutagen.

In long-term studies, orally administered styrene increased the incidence of lung tumours in mice at high dose levels but had no carcinogenic effect in rats. Styrene-7,8-oxide was carcinogenic in rats after oral administration. IARC has classified styrene in Group 2B. The available data suggest that the carcinogenicity of styrene is due to overloading of the detoxification mechanism for styrene-7,8-oxide (e.g., glutathione depletion).

A TDI of 7.7 µg/kg of body weight was derived using a NOAEL of 7.7 mg/kg of body weight per day in a 2-year drinking-water study in rats and applying an uncertainty factor of 1000 (100 for intra- and interspecies variation and 10 for carcinogenicity and genotoxicity of the reactive intermediate styrene-7,8-oxide). This yields a guideline value of 20 µg/litre (rounded figure), allocating 10% of the TDI to drinking-water. It should be noted that styrene may affect the acceptability of drinking-water at this concentration (see page 128).

Polynuclear aromatic hydrocarbons
A large number of polynuclear aromatic hydrocarbons (PAHs) from a variety of combustion and pyrolysis sources have been identified in the environment. The main source of human exposure to PAHs is food, with drinking-water contributing only minor amounts.

Little information is available on the oral toxicity of PAHs, especially after long-term exposure. Benzo[*a*]pyrene, which constitutes a minor fraction of total PAHs, has been found to be carcinogenic in mice by the oral route of administration; some PAH compounds have been found to be carcinogenic by non-oral routes, and others have been determined to have a low potential for carcinogenicity. Benzo[*a*]pyrene has been found to be mutagenic in a number of *in vitro* and *in vivo* assays.

Adequate data upon which to base a quantitative assessment of the carcinogenicity of ingested PAHs are available only for benzo[*a*]pyrene, which appears to be a local carcinogen in that it induces tumours at the site of administration. Administration of benzo[*a*]pyrene in the diet of mice resulted in an increased incidence of forestomach tumours. Owing to the unusual protocol followed in this study, which involved variable dosing patterns and age of sacrifice, these data could not be accurately extrapolated using the linearized multistage model normally applied in the derivation of these drinking-water guidelines. However, a quantitative risk assessment was conducted using the two-stage birth–death mutation model. The resulting guideline value for benzo[*a*]pyrene in drinking-water, corresponding to an excess lifetime cancer risk of $10^{-5}$, is 0.7 µg/litre.

There are insufficient data available to derive drinking-water guidelines for other PAHs. However, the following recommendations are made for the PAH group:

- Because of the close association of PAHs with suspended solids, the application of treatment, when necessary, to achieve the recommended level of turbidity will ensure that PAH levels are reduced to a minimum.
- Contamination of water with PAHs should not occur during water treatment or distribution. Therefore, the use of coal-tar-based and similar materials for pipe linings and coatings on storage tanks should be discontinued. It is recognized that it may be impracticable to remove coal-tar linings from existing pipes. However, research into methods of minimizing the leaching of PAHs from such lining materials should be carried out.
- To monitor PAH levels, the use of several specific compounds as indicators for the group as a whole is recommended. The choice of indicator compounds will vary for each individual situation. PAH levels should be monitored regularly in order to determine the background levels against which any changes can be assessed so that remedial action can be taken, if necessary.
- In situations where contamination of drinking-water by PAHs has occurred, the specific compounds present and the source of the contamination should be identified, as the carcinogenic potential of PAH compounds varies.

### *Chlorinated benzenes*

Monochlorobenzene

Releases of monochlorobenzene (MCB) to the environment are thought to be mainly due to volatilization losses associated with its use as a solvent in pesticide formulations, as a degreasing agent, and from other industrial applications. The major source of human exposure is probably air.

MCB is of low acute toxicity. Oral exposure to high doses of MCB affects mainly the liver, kidneys, and haematopoietic system. There is limited evidence of carcinogenicity in male rats, with high doses increasing the occurrence of neoplastic nodules in the liver. The majority of evidence suggests that MCB is not mutagenic; although it binds to DNA *in vivo*, the level of binding is low.

# 3. CHEMICAL ASPECTS

A TDI of 85.7 µg/kg of body weight was calculated by applying an uncertainty factor of 500 (100 for inter- and intraspecies variation and 5 for the limited evidence of carcinogenicity) to a NOAEL of 60 mg/kg of body weight for neoplastic nodules identified in a 2-year rat study with 5 days per week dosing by gavage. This gives a guideline value of 300 µg/litre (rounded figure) based on an allocation of 10% of the TDI to drinking-water. However, this value far exceeds the lowest reported taste and odour threshold for MCB in water (see page 129).

## Dichlorobenzenes

The dichlorobenzenes (DCBs) are widely used in industry and in domestic products such as odour-masking agents, chemical dyestuffs, and pesticides. Sources of human exposure are predominantly air and food.

### *1,2-Dichlorobenzene*

1,2-DCB is of low acute toxicity by the oral route of exposure. Oral exposure to high doses of 1,2-DCB affects mainly the liver and kidneys. The balance of evidence suggests that 1,2-DCB is not genotoxic, and there is no evidence for its carcinogenicity in rodents.

A TDI of 429 µg/kg of body weight was calculated for 1,2-DCB by applying an uncertainty factor of 100 (for inter- and intraspecies variation) to a NOAEL of 60 mg/kg of body weight per day for tubular degeneration of the kidney identified in a 2-year mouse gavage study with administration 5 days per week. This gives a guideline value of 1000 µg/litre (rounded figure) based on an allocation of 10% of the TDI to drinking-water. This value far exceeds the lowest reported taste threshold of 1,2-DCB in water (see page 129).

### *1,3-Dichlorobenzene*

There are insufficient toxicological data on this compound to permit a guideline value to be proposed, but it should be noted that it is rarely found in drinking-water.

### *1,4-Dichlorobenzene*

1,4-DCB is of low acute toxicity, but there is evidence that it increases the incidence of renal tumours in rats and of hepatocellular adenomas and carcinomas in mice after long-term exposure. IARC has placed 1,4-DCB in Group 2B.

1,4-DCB is not considered to be genotoxic, and the relevance for humans of the tumours observed in animals is doubtful. It is therefore valid to calculate a guideline value using the TDI approach. A TDI of 107 µg/kg of body weight was calculated by applying an uncertainty factor of 1000 (100 for inter- and intraspecies variation and 10 because a LOAEL was used instead of a NOAEL and because the toxic end-point is carcinogenicity) to a LOAEL of 150 mg/kg of body weight per day for kidney effects identified in a 2-year rat study (administration 5 days per week). A guideline value of 300 µg/litre (rounded figure) is proposed based on an allocation of 10% of the TDI to drinking-water. This value far exceeds the lowest reported odour threshold of 1,4-DCB in water (see page 129).

Trichlorobenzenes
Releases of trichlorobenzenes (TCBs) into the environment occur through their manufacture and use as industrial chemicals, chemical intermediates, and solvents. TCBs are found in drinking-water but rarely at levels above 1 µg/litre. General population exposure will primarily result from air and food.

The TCBs are of moderate acute toxicity. After short-term oral exposure, all three isomers show similar toxic effects, predominantly on the liver. Long-term toxicity and carcinogenicity studies via the oral route have not been carried out, but the data available suggest that all three isomers are non-genotoxic.

A TDI of 7.7 µg/kg of body weight was calculated by applying an uncertainty factor of 1000 (100 for inter- and intraspecies variation and 10 for the short duration of the study) to the NOAEL of 7.7 mg/kg of body weight per day for liver toxicity identified in a 13-week rat study. The guideline value would be 20 µg/litre (rounded figure) for each isomer based on an allocation of 10% of the TDI to drinking-water; however, because of the similarity in the toxicity of the TCB isomers, a guideline value of 20 µg/litre is proposed for total TCBs. This value exceeds the lowest reported odour threshold in water (see page 129).

## *Miscellaneous organic constituents*

Di(2-ethylhexyl)adipate
Di(2-ethylhexyl)adipate (DEHA) is used mainly as a plasticizer for synthetic resins such as polyvinyl chloride (PVC). As a consequence of its use in PVC films, food is the most important source of human exposure (up to 20 mg/day). Reports of the presence of DEHA in surface water and drinking-water are scarce, but DEHA has occasionally been identified in drinking-water at levels of a few micrograms per litre.

DEHA is of low short-term toxicity; however, dietary levels above 6000 mg/kg of feed induce peroxisomal proliferation in the liver of rodents. This effect is often associated with the development of liver tumours. DEHA induced liver carcinomas in female mice at very high doses but not in male mice or rats. It is not genotoxic. IARC has placed DEHA in Group 3.

Although DEHA is carcinogenic in mice, the toxicity profile and lack of mutagenicity of DEHA support the use of a TDI approach to setting a guideline value for DEHA in drinking-water. A TDI of 280 µg/kg of body weight was calculated by applying an uncertainty factor of 100 (for inter- and intraspecies variation) to the lowest NOAEL for DEHA of 28 mg/kg of body weight per day based on fetotoxicity in rats. The guideline value is 80 µg/litre (rounded figure) based on an allocation of 1% of the TDI to drinking-water.

### Di(2-ethylhexyl)phthalate

Di(2-ethylhexyl)phthalate (DEHP) is used primarily as a plasticizer. It is found in surface water, ground water, and drinking-water in concentrations of a few micrograms per litre. In polluted surface and ground water, concentrations of hundreds of micrograms per litre have been reported.

The reliability of some data on environmental water samples is questionable because of secondary contamination during sampling and working-up procedures. Concentrations that exceed the solubility more than 10-fold have been reported.

Exposure among individuals may vary considerably because of the broad nature of products into which DEHP is incorporated. In general, food will be the main exposure route.

In rats, DEHP is readily absorbed from the gastrointestinal tract. In primates (including humans), absorption after ingestion is lower. Species differences are also observed in the metabolic profile. Most species excrete primarily the conjugated mono-ester in urine. Rats, however, predominantly excrete terminal oxidation products. DEHP is widely distributed in the body, with highest levels in liver and adipose tissue, without showing significant accumulation.

The acute oral toxicity is low. The most striking effect in short-term toxicity studies is the proliferation of hepatic peroxisomes, indicated by increased peroxisomal enzyme activity and histopathological changes. The available information suggests that primates, including humans, are far less sensitive to this effect than rodents.

In long-term oral carcinogenicity studies, hepatocellular carcinomas were found in rats and mice. IARC has concluded that DEHP is possibly carcinogenic to humans (Group 2B). In 1988, JECFA evaluated DEHP and recommended that human exposure to this compound in food be reduced to the lowest level attainable. The Committee considered that this might be achieved by using alternative plasticizers or alternatives to plastic material containing DEHP.

In a variety of *in vitro* and *in vivo* studies, DEHP and its metabolites have shown no evidence of genotoxicity, with the exception of induction of aneuploidy and cell transformation.

Based on the absence of evidence for genotoxicity and the suggested relationship between prolonged proliferation of liver peroxisomes and the occurrence of hepatocellular carcinomas, a TDI was derived using the lowest observed NOAEL of 2.5 mg/kg of body weight per day based on peroxisomal proliferation in the liver in rats. Although the mechanism for hepatocellular tumour induction is not fully resolved, use of a NOAEL derived from the species by far the most sensitive with respect to the particularly sensitive end-point of peroxisomal proliferation justifies the use of an uncertainty factor of 100 (for inter- and intraspecies variation). Consequently, the TDI is 25 µg/kg of body weight. This yields a guideline value of 8 µg/litre (rounded figure), allocating 1% of the TDI to drinking-water.

Acrylamide
Residual acrylamide monomer occurs in polyacrylamide coagulants used in the treatment of drinking-water. In general, the maximum authorized dose of polymer is 1 mg/litre. At a monomer content of 0.05%, this corresponds to a maximum theoretical concentration of 0.5 µg/litre of the monomer in water. Practical concentrations may be lower by a factor of two to three. This applies to the anionic and nonionic polyacrylamides, but residual levels from cationic polyacrylamides may be higher. Polyacrylamides are also used as grouting agents in the construction of drinking-water reservoirs and wells. Additional human exposure might result from food, owing to the use of polyacrylamide in food processing.

Following ingestion, acrylamide is readily absorbed from the gastrointestinal tract and widely distributed in body fluids. Acrylamide can cross the placenta. It is neurotoxic, affects germ cells, and impairs reproductive function.

In mutagenicity assays, acrylamide was negative in the Ames test but induced gene mutations in mammalian cells and chromosomal aberrations *in vitro* and *in vivo*. In a long-term carcinogenicity study in rats exposed via drinking-water, acrylamide induced scrotal, thyroid, and adrenal tumours in males, and mammary, thyroid, and uterine tumours in females. IARC has placed acrylamide in Group 2B.

On the basis of the available information, it was concluded that acrylamide is a genotoxic carcinogen. Therefore, the risk evaluation was carried out using a non-threshold approach.

On the basis of combined mammary, thyroid, and uterine tumours observed in female rats in a drinking-water study, and using the linearized multistage model, a guideline value associated with an excess lifetime cancer risk of $10^{-5}$ is estimated to be 0.5 µg/litre.

The most important source of drinking-water contamination by acrylamide is the use of polyacrylamide flocculants that contain residual acrylamide monomer. Although the practical quantification level for acrylamide is generally in the order of 1 µg/litre, concentrations in drinking-water can be controlled by product and dose specification.

Epichlorohydrin
Epichlorohydrin (ECH) is used for the manufacture of glycerol, unmodified epoxy resins, and water-treatment resins. No quantitative data are available on its

occurrence in food or drinking-water. ECH is hydrolysed in aqueous media.

ECH is rapidly and extensively absorbed following oral, inhalation or dermal exposure. It binds easily to cellular components.

Major toxic effects are local irritation and damage to the central nervous system. It induces squamous cell carcinomas in the nasal cavity by inhalation and forestomach tumours by the oral route. It has been shown to be genotoxic *in vitro* and *in vivo*. IARC has placed ECH in Group 2A.

Although ECH is a genotoxic carcinogen, the use of the linearized multistage model for estimating cancer risk was considered inappropriate because tumours are seen only at the site of administration, where ECH is highly irritating.

A TDI of 0.143 µg/kg of body weight was therefore calculated by applying an uncertainty factor of 10 000 (100 for inter- and intraspecies variation, 10 for the use of a LOAEL instead of a NOAEL, and 10 reflecting carcinogenicity) to a LOAEL of 2 mg/kg of body weight per day for forestomach hyperplasia in a 2-year study in rats by gavage (administration 5 days per week). This gives a provisional guideline value of 0.4 µg/litre (rounded figure) based on an allocation of 10% of the TDI to drinking-water. A practical quantification level for ECH is of the order of 30 µg/litre, but concentrations in drinking-water can be controlled by specifying the ECH content of products coming into contact with it.

Hexachlorobutadiene
Hexachlorobutadiene (HCBD) is used as a solvent in chlorine gas production, a pesticide, an intermediate in the manufacture of rubber compounds, and a lubricant. Concentrations of up to 6 µg/litre have been reported in the effluents from chemical manufacturing plants. It is also found in air and food.

HCBD is easily absorbed and metabolized via conjugation with glutathione. This conjugate can be further metabolized to a nephrotoxic derivative.

Kidney tumours were observed in a long-term oral study in rats. HCBD has not been shown to be carcinogenic by other routes of exposure. IARC has placed HCBD in Group 3. Positive and negative results for HCBD have been obtained in bacterial assays for point mutation; however, several metabolites have given positive results.

On the basis of the available metabolic and toxicological information, it was considered that a TDI approach was most appropriate for derivation of a guideline value. A TDI of 0.2 µg/kg of body weight was therefore calculated by applying an uncertainty factor of 1000 (100 for inter- and intraspecies variation and 10 for limited evidence of carcinogenicity and the genotoxicity of some metabolites) to the NOAEL of 0.2 mg/kg of body weight per day for renal toxicity in a 2-year feeding study in rats. This gives a guideline value of 0.6 µg/litre, based on an allocation of 10% of the TDI to drinking-water. A practical quantification level for HCBD is of the order of 2 µg/litre, but concentrations in drinking-water can be controlled by specifying the HCBD content of products coming into contact with it.

### Edetic acid

Edetic acid (ethylenediaminetetraacetic acid; EDTA) and its salts are used in many industrial processes, in domestic products, and as food additives. EDTA is also used as a drug in chelation therapy. It is poorly degraded, and there are substantial releases to the aquatic environment. Levels in natural water of up to 0.9 mg/litre have been recorded but are usually less than 0.1 mg/litre.

The toxicology database on EDTA is relatively old, and studies in laboratory animals are complicated by the fact that EDTA forms complexes with zinc in the gastrointestinal tract. EDTA is poorly absorbed and is considered to be of low toxicity. There is no information on mutagenicity and only limited data on carcinogenicity. In 1973, JECFA proposed an ADI for calcium disodium edetate as a food additive of 2.5 mg/kg of body weight (1.9 mg/kg of body weight as the free acid). However, JECFA recommended that no sodium edetate should remain in food.

An extra uncertainty factor of 10 was introduced to reflect the fact that the JECFA ADI has not been considered since 1973 and concern over zinc complexation, giving a TDI of 190 µg/kg of body weight. In view of the possibility of zinc complexation, a provisional guideline value was derived assuming consumption of 1 litre of water by a 10-kg child. The provisional guideline value is thus 200 µg/litre (rounded figure), allocating 10% of the TDI to drinking-water.

### Nitrilotriacetic acid

Nitrilotriacetic acid (NTA) is used primarily in laundry detergents as a replacement for phosphates and in the treatment of boiler water to prevent accumulation of mineral scale. Concentrations in drinking-water usually do not exceed a few micrograms per litre.

NTA is not metabolized in animals and is rapidly eliminated, although some may be briefly retained in bone. It is of low acute toxicity to animals, but it has been shown to produce kidney tumours in rodents following long-term exposure to high doses. IARC has placed NTA in Group 2B. It is not genotoxic, and the reported induction of tumours is believed to be due to cytotoxicity resulting from the chelation of divalent cations such as zinc and calcium in the urinary tract, leading to the development of hyperplasia and subsequently neoplasia.

Because NTA is non-genotoxic and induces tumours only after prolonged exposure to doses higher than those that produce nephrotoxicity, the guideline value was determined using a TDI approach. A TDI of 10 µg/kg of body weight was calculated by applying an uncertainty factor of 1000 (100 for inter- and intraspecies variation and 10 for carcinogenic potential at high doses) to the NOAEL of 10 mg/kg of body weight per day for nephritis and nephrosis in a 2-year study in rats. Because there is no substantial exposure from other sources, 50% of the TDI was allocated to drinking-water, resulting in a guideline value of 200 µg/litre (rounded figure).

Organotins

The group of chemicals known as the organotins is composed of a large number of compounds with differing properties and applications. The most widely used of the organotins are the disubstituted compounds, which are employed as stabilizers in plastics, including polyvinyl chloride (PVC) water pipes, and the trisubstituted compounds, which are widely used as biocides.

*Dialkyltins*

The disubstituted compounds that may leach from PVC water pipes for a short time after installation are primarily immunotoxins, although they appear to be of low general toxicity. The data available are insufficient to permit the proposal of guideline values for individual dialkyltins.

*Tributyltin oxide*

Tributyltin oxide (TBTO) is widely used as a biocide in wood preservatives and antifouling paints. It is extremely toxic to aquatic life, and its use is being reduced in some countries. There are only limited exposure data; however, exposure from food, except from certain seafoods, is unlikely.

TBTO is not genotoxic. One carcinogenicity study has been reported in which neoplastic changes were observed in endocrine organs, but the significance of these changes is considered questionable. The most sensitive end-point appears to be immunotoxicity, with a lowest NOAEL of 0.025 mg/kg of body weight per day in a 17-month feeding study in rats related to suppression of resistance to the nematode *Trichinella spiralis*. The significance to humans of this finding is not completely clear, but this NOAEL is consistent, within an order of magnitude, with other NOAELs for long-term toxicity.

A TDI of 0.25 µg/kg of body weight was calculated by applying an uncertainty factor of 100 (for inter- and intraspecies variation) to the NOAEL of 0.025 mg/kg of body weight per day for suppression of resistance to *T. spiralis*. The guideline value for TBTO is 2 µg/litre (rounded figure) based on an allocation of 20% of the TDI to drinking-water.

The database on the toxicity of the other trisubstituted organotin compounds is either limited or rather old. It was therefore not considered appropriate to propose guideline values for these compounds.

### 3.6.3 Pesticides

It is recognized that the degradation products of pesticides may be a problem in drinking-water. In most cases, however, the toxicities of these degradation products have not been taken into consideration in these guidelines, as there are inadequate data on their identity, presence, and biological activity.

### Alachlor

Alachlor is a pre- and post-emergence herbicide used to control annual grasses and many broad-leaved weeds in maize and a number of other crops. It is lost from soil mainly through volatilization, photodegradation, and biodegradation. Many alachlor degradation products have been identified in soil. Alachlor has been detected in ground and surface water. It has also been detected in drinking-water at levels below 2 µg/litre.

On the basis of available experimental data, evidence for the genotoxicity of alachlor is considered to be equivocal. However, a metabolite of alachlor has been shown to be mutagenic. Available data from two studies in rats clearly indicate that alachlor is carcinogenic, causing benign and malignant tumours of the nasal turbinate, malignant stomach tumours, and benign thyroid tumours.

In view of the data on carcinogenicity, a guideline value was calculated by applying the linearized multistage model to data on the incidence of nasal tumours in rats. The guideline value in drinking-water, corresponding to an excess lifetime cancer risk of $10^{-5}$, is 20 µg/litre.

### Aldicarb

Aldicarb is a systemic pesticide used to control nematodes in soil and insects and mites on a variety of crops. It is very soluble in water and is highly mobile in soil. It degrades mainly by biodegradation and hydrolysis, persisting for weeks to months. It has been frequently found as a contaminant in ground water.

Aldicarb is one of the most acutely toxic pesticides in use, although the only consistently observed toxic effect with both long-term and single-dose administration is acetylcholinesterase inhibition. It is metabolized to the sulfoxide and sulfone.

The weight of evidence indicates that aldicarb is not genotoxic or carcinogenic. IARC has concluded that aldicarb is not classifiable as to its carcinogenicity (Group 3).

For the purposes of deriving a guideline for drinking-water, a 29-day study in rats was used, in which a 1:1 mixture of aldicarb sulfoxide and aldicarb sulfone (the ratio most commonly found in drinking-water) was administered in drinking-water. The NOAEL was 0.4 mg/kg of body weight per day based on acetylcholinesterase inhibition. An uncertainty factor of 100 (for inter- and intraspecies variation) was applied, giving a TDI of 4 µg/kg of body weight. No allowance was made for the short duration of the study in view of the extremely sensitive and rapidly reversible biological end-point used. The guideline value is 10 µg/litre (rounded figure), assuming an allocation of 10% of the TDI to drinking-water.

### Aldrin and dieldrin

Aldrin and dieldrin are chlorinated pesticides that are used against soil-dwelling pests, for wood protection, and, in the case of dieldrin, against insects of public health importance. The two compounds are closely related with respect to their toxicology and mode of action. Aldrin is rapidly converted to dieldrin under most environmental conditions and in the body. Dieldrin is a highly persistent organochlorine compound that has low mobility in soil and can be lost to the atmosphere. It is occasionally found in water. Dietary exposure to aldrin/dieldrin is very low and decreasing. Since the early 1970s, a number of countries have either severely restricted or banned the use of both compounds, particularly in agriculture.

Both compounds are highly toxic in experimental animals, and cases of poisoning in humans have occurred. Aldrin and dieldrin have more than one mechanism of toxicity. The target organs are the central nervous system and the liver. In long-term studies, dieldrin was shown to produce liver tumours in both sexes of two strains of mice. It did not produce an increase in tumours in rats and does not appear to be genotoxic.

IARC has classified aldrin and dieldrin in Group 3. It is considered that all the available information on aldrin and dieldrin taken together, including studies on humans, supports the view that, for practical purposes, these chemicals make very little contribution, if any, to the incidence of cancer in humans. Therefore, a TDI approach can be used to calculate a guideline value.

In 1977, JMPR recommended an ADI of 0.1 µg/kg of body weight (combined total for aldrin and dieldrin). This was based on NOAELs of 1 mg/kg of diet in the dog and 0.5 mg/kg of diet in the rat, which are equivalent to 0.025 mg/kg of body weight per day in both species. JMPR applied an uncertainty factor of 250 based on concern about carcinogenicity observed in mice.

This ADI is reaffirmed. Although the levels of aldrin/dieldrin in food have been decreasing, dieldrin is highly persistent and accumulates in body tissues. There is also potential for exposure from the atmosphere of houses where it is used for termite control. The guideline value is therefore based on an allocation of 1% of the ADI to drinking-water, giving a value of 0.03 µg/litre.

### Atrazine

Atrazine is a selective pre- and early post-emergence herbicide. It has been found in surface and ground water as a result of its mobility in soil. It is relatively stable in soil and aquatic environments, with a half-life measured in months, but is degraded by photolysis and microbial degradation in soil

The weight of evidence from a wide variety of genotoxicity assays indicates that atrazine is not genotoxic. There is evidence that atrazine can induce mammary tumours in rats. It is highly probable that the mechanism for this process is non-genotoxic. No significant increase in neoplasia has been observed in mice. IARC has concluded that there is inadequate evidence in humans and limited evidence in experimental animals for the carcinogenicity of atrazine (Group 2B).

A TDI approach can therefore be used to calculate a guideline value. Based on a NOAEL of 0.5 mg/kg of body weight per day in a carcinogenicity study in the rat and an uncertainty factor of 1000 (100 for inter- and intraspecies variation and 10 to reflect potential neoplasia), a TDI of 0.5 µg/kg of body weight was calculated. With an allocation of 10% of the TDI to drinking-water, the guideline value is 2 µg/litre (rounded figure).

Bentazone
Bentazone is a broad-spectrum herbicide used for a variety of crops. It photodegrades in soil and water but is very mobile in soil and is moderately persistent in the environment. It has been found in ground water and has a high affinity for the water compartment.

Long-term studies conducted in rats and mice have not indicated a carcinogenic potential, and a variety of *in vitro* and *in vivo* assays have indicated that bentazone is not genotoxic.

JMPR evaluated bentazone in 1991 and established an ADI of 0.1 mg/kg of body weight by applying an uncertainty factor of 100 to a NOAEL of 10 mg/kg of body weight per day, based upon haematological effects at higher doses, derived from a 2-year dietary study in rats and supported by NOAELs in mice and dogs. To allow for uncertainties regarding dietary exposure, 1% of the ADI was allocated to drinking-water, resulting in a guideline value of 30 µg/litre.

Carbofuran
Carbofuran is a systemic acaricide, insecticide, and nematocide. It can undergo photodegradation or chemical and microbial degradation. It is sufficiently mobile and persistent to leach from soil, and it has been found in ground water at typical levels of 1–5 µg/litre.

From a 1-year study in dogs, a NOAEL of 0.5 mg/kg of body weight per day was derived. The NOAEL for systemic effects in dams in a rat teratology study was 0.1 mg/kg of body weight per day. On the basis of the available studies, carbofuran does not appear to be carcinogenic or genotoxic.

The clinical manifestations of carbofuran poisoning resemble those of organophosphorus intoxication. The available data on humans show that, whereas clinical signs of acetylcholinesterase inhibition were observed after a single oral dose of 0.10 mg/kg of body weight, they were absent at 0.05 mg/kg of body weight. Hence, this latter level can be regarded as a NOAEL in humans.

A TDI of 1.67 µg/kg of body weight was calculated by applying an uncertainty factor of 30 (10 for intraspecies variation and 3 for the steep dose-response curve) to the NOAEL of 0.05 mg/kg of body weight in humans. This TDI is supported by observations in laboratory animals, giving an adequate margin of safety for the NOAELs in rat and dog. An allocation of 10% of the TDI to drinking-water results in the guideline value of 5 µg/litre (rounded figure).

## Chlordane

Chlordane is a broad-spectrum insecticide that has been used since 1947. Its use has recently been increasingly restricted in many countries, and it is now used mainly to destroy termites by subsurface injection into soil.

Chlordane is a mixture of stereoisomers, with the *cis* and *trans* forms predominating. It is very resistant to degradation, is highly immobile in soil, and migrates very poorly to ground water, where it has only rarely been found. It is readily lost to the atmosphere.

In experimental animals, prolonged exposure in the diet causes liver damage. Chlordane produces liver tumours in mice, but the weight of evidence indicates that it is not genotoxic. Chlordane can interfere with cell communication *in vitro*, a characteristic of many tumour promoters.

IARC re-evaluated chlordane in 1991 and concluded that there is inadequate evidence for its carcinogenicity in humans and sufficient evidence for its carcinogenicity in animals, classifying it in Group 2B.

JMPR re-reviewed chlordane in 1986 and established an ADI of 0.5 µg/kg of body weight by applying an uncertainty factor of 100 to the NOAEL of 0.05 mg/kg of body weight per day derived from a long-term dietary study in rats.

Although levels of chlordane in food have been decreasing, it is highly persistent and has a high bioaccumulation potential. An allocation of 1% of the JMPR ADI to drinking-water gives a guideline value of 0.2 µg/litre (rounded figure).

## Chlorotoluron

Chlorotoluron is a pre- or early post-emergence herbicide that is slowly biodegradable and mobile in soil. It has been detected in drinking-water at concentrations of less than 1 µg/litre. There is only very limited exposure to this compound from food.

Chlorotoluron is of low toxicity in acute, short-term, and long-term exposures in animals, but it has been shown to cause an increase in adenomas and carcinomas of the kidneys of male mice given high doses for 2 years. Chlorotoluron and its metabolites have shown no evidence of genotoxicity.

In view of this, the guideline value for chlorotoluron was calculated using a TDI approach. An uncertainty factor of 1000 (100 for inter- and intraspecies variation and 10 for evidence of carcinogenicity) was applied to the NOAEL of

11.3 mg/kg of body weight per day in a 2-year feeding study in mice to give a TDI of 11.3 µg/kg of body weight. An allocation of 10% of the TDI to drinking-water results in the guideline value of 30 µg/litre (rounded figure).

## DDT

The structure of DDT allows several different isomeric forms, and commercial products consist predominantly of $p,p'$-DDT. In some countries the use of DDT has been restricted or even prohibited, but it is still extensively used elsewhere, both in agriculture and for vector control. It is a persistent insecticide, stable under most environmental conditions; DDT and some of its metabolites are resistant to complete breakdown by soil microorganisms.

In small doses, DDT and its metabolites are almost totally absorbed in humans following ingestion or inhalation and are stored in adipose tissue and milk.

IARC has concluded that there is insufficient evidence in humans and sufficient evidence in experimental animals for the carcinogenicity of DDT (Group 2B) based upon liver tumours observed in rats and mice. Moreover, JMPR has concluded that the mouse is particularly sensitive to DDT because of its genetic and metabolic characteristics. In most studies, DDT did not induce genotoxic effects in rodent or human cell systems, nor was it mutagenic in fungi or bacteria. DDT impaired reproduction in several species.

A guideline value was derived from the ADI of 0.02 mg/kg of body weight recommended by JMPR in 1984, based on NOAELs of 6.25 mg/kg of body weight per day in rats, 10 mg/kg of body weight per day in monkeys, and 0.25 mg/kg of body weight per day in humans. For adults, this ADI would provide a 500-fold margin of safety for the NOAEL of 10 mg/kg of body weight per day found in the study in monkeys.

Because infants and children may be exposed to greater amounts of chemicals in relation to their body weight, and because of concern over the bioaccumulation of DDT, the guideline value was calculated on the basis of a 10-kg child drinking 1 litre of water per day. Moreover, because there is significant exposure to DDT by routes other than water, a 1% allocation of the ADI to drinking-water was chosen. This leads to a guideline value for DDT and its metabolites in drinking-water of 2 µg/litre.

This guideline value exceeds the water solubility of DDT of 1 µg/litre. However, some DDT may be adsorbed onto the small amount of particulate matter present in drinking-water, so that the guideline value of 2 µg/litre could be reached under certain circumstances.

It should be emphazised that, as for all pesticides, the recommended guideline value for DDT in drinking-water is set at a level to protect human health; it may not be suitable for the protection of the environment or aquatic life. The benefits of DDT use in malaria and other vector control programmes far outweigh any health risk from the presence of DDT in drinking-water.

## 1,2-Dibromo-3-chloropropane

1,2-Dibromo-3-chloropropane (DBCP) is a soil fumigant that is highly soluble in water. It has a taste and odour threshold in water of 10 µg/litre. A limited survey found DBCP at levels of up to a few micrograms per litre in drinking-water. DBCP was also detected in vegetables grown in treated soils, and low levels have been detected in air.

On the basis of animal data from different strains of rats and mice, DBCP was determined to be carcinogenic in both sexes by the oral, inhalation, and dermal routes. DBCP was also determined to be a reproductive toxicant in humans and several species of laboratory animals. DBCP was found to be genotoxic in a majority of *in vitro* and *in vivo* assays. IARC has classified DBCP in Group 2B based upon sufficient evidence of carcinogenicity in animals. Recent epidemiological evidence suggests an increase in cancer mortality in individuals exposed to high levels of DBCP.

The linearized multistage model was applied to the data on the incidence of stomach, kidney, and liver tumours in the male rat in a 104-week dietary study. The concentration in drinking-water relating to an excess lifetime cancer risk of $10^{-5}$ is 1 µg/litre. The guideline value of 1 µg/litre should be protective for the reproductive toxicity of DBCP. For a contaminated water supply, extensive treatment (e.g., air stripping followed by adsorption to granular activated carbon) would be required to reduce the level of DBCP to the guideline value.

## 2,4-Dichlorophenoxyacetic acid (2,4-D)

2,4-D is a chlorophenoxy herbicide that is used extensively in the control of broad-leaved weeds. The half-life for biodegradation in soil ranges from a few days to 6 weeks, while the half-life in water ranges from one to several weeks. Limited monitoring data indicate that levels in drinking-water generally do not exceed a few micrograms per litre. 2,4-D is rarely found in foods.

IARC has classified chlorophenoxy herbicides in Group 2B. Although in one study in humans there was a marginally significant trend in the excess risk of non-Hodgkin lymphoma with increasing duration of exposure to chlorophenoxy herbicides, it is not possible to evaluate the carcinogenic potential of 2,4-D *per se* on the basis of available epidemiological data. A dose-related increase in the incidence of astrocytomas of the brain was reported in a carcinogenicity study in rats. However, this study was considered to be of limited value for the evaluation of carcinogenicity. 2,4-D was found to be non-genotoxic in the limited number of studies conducted.

Because the data on the carcinogenic potential of 2,4-D are inadequate, and because 2,4-D has not been found to be genotoxic, the guideline value was derived using a TDI approach for other toxic end-points. The NOAEL for effects on the kidney in 2-year studies in rats and mice was considered to be 1 mg/kg of body weight per day. An uncertainty factor of 100 (for intra- and interspecies variation) was applied to this NOAEL, resulting in a TDI of 10 µg/kg of body weight. The use of an additional uncertainty factor for carcinogenicity was considered unnecessary, as this NOAEL should provide a sufficient margin of safety with respect to the lowest dose that was associated with an increase in brain tumours in rats. The guideline value, based on an allocation of 10% of the TDI to drinking-water, is 30 µg/litre.

### 1,2-Dichloropropane

1,2-Dichloropropane, also known as propylene dichloride, is used primarily as a chemical intermediate, lead scavenger for antiknock fluids, dry-cleaning and metal-degreasing solvent, and soil fumigant. Because of its solubility and in spite of its high vapour pressure, it can contaminate water.

There is a relatively limited database on the toxicity of 1,2-dichloropropane, but it was shown to be a mutagen in some short-term assays *in vitro*.

When administered orally, 1,2-dichloropropane produced statistically significant increases in the incidence of hepatocellular adenomas and carcinomas in both sexes of mice. There was marginal evidence of carcinogenicity in female rats. IARC has classified 1,2-dichloropropane in Group 3.

A guideline value was derived using a TDI approach. A LOAEL of 100 mg/kg of body weight per day was identified on the basis of a variety of systemic effects in a 13-week oral study in rats (administration 5 days per week). A TDI of 7.14 µg/kg of body weight was calculated by applying an uncertainty factor of 10 000 (100 for inter- and intraspecies variation, 10 because a LOAEL was used instead of a NOAEL, and 10 to reflect limited evidence of carcinogenicity in animals and a limited toxicity database, particularly for reproductive studies). With an allocation of 10% of the TDI to drinking-water, the provisional guideline value is 20 µg/litre (rounded figure).

### 1,3-Dichloropropane

1,3-Dichloropropane has several industrial uses and may be found as a contaminant of soil fumigants containing 1,3-dichloropropene. 1,3-Dichloropropane is rarely found in water. It is of low acute toxicity. There is some indication that it may be genotoxic in bacterial systems. No short-term, long-term, reproductive, or developmental toxicity data pertinent to exposure via drinking-water could be located in the literature. The available data were considered insufficient to permit recommendation of a guideline value.

### 1,3-Dichloropropene

1,3-Dichloropropene is a soil fumigant, the commercial product being a mixture of *cis* and *trans* isomers. It is used to control a wide variety of soil pests, particularly nematodes in sandy soils. Notwithstanding its high vapour pressure, it is soluble in water at the gram per litre level and can be considered a potential water contaminant.

1,3-Dichloropropene is a direct-acting mutagen that has been shown to produce forestomach tumours following long-term oral gavage exposure in rats and mice. Tumours have also been found in the bladder and lungs of female mice and the liver of male rats. Long-term inhalation studies in the rat have proved negative, whereas in inhalation studies in mice some benign lung tumours have been reported. IARC has classified 1,3-dichloropropene in Group 2B.

Based on observation of lung and bladder tumours in female mice in a 2-year gavage study and using the linearized multistage model, a guideline value corresponding to an excess lifetime cancer risk of $10^{-5}$ is estimated to be 20 $\mu$g/litre.

Ethylene dibromide (EDB)
EDB, also known as 1,2-dibromoethane, is used as an active additive in leaded petrol, an insecticidal fumigant, and an industrial chemical.

EDB is photodegradable with a short persistence in air; however, it can persist for much longer in other environmental compartments. It is volatile, but its solubility and its resistance to degradation make this chemical a potential contaminant of ground water.

EDB is a bifunctional alkylating agent that induces a variety of effects, including male reproductive effects. IARC re-evaluated the data on EDB in 1987 and concluded that the evidence for carcinogenicity to humans was inadequate but that the animal data were sufficient to establish carcinogenicity, assigning EDB to Group 2A. EDB has been found to be genotoxic in both *in vitro* and *in vivo* assays.

Although EDB appears to be a genotoxic carcinogen, the studies to date are inadequate for mathematical risk extrapolation. Therefore, a guideline value for EDB has not been derived. EDB should be re-evaluated as soon as new data become available.

Heptachlor and heptachlor epoxide
Heptachlor is a broad-spectrum insecticide, the use of which has been banned or restricted in many countries. At present, the major use of heptachlor is for termite control by subsurface injection into soil.

Heptachlor is quite persistent in soil, where it is mainly transformed to its epoxide. Heptachlor epoxide is very resistant to further degradation. Heptachlor and heptachlor epoxide bind to soil particles and migrate very slowly. Heptachlor and heptachlor epoxide have been found in drinking-water at levels of nanograms per litre. Diet is considered to represent the major source of exposure to heptachlor, although intake is decreasing.

Prolonged exposure to heptachlor has been associated with damage to the liver and central nervous system toxicity.

In 1991, IARC reviewed the data on heptachlor and concluded that the evidence for carcinogenicity was sufficient in animals and inadequate in humans, classifying it in Group 2B.

JMPR has evaluated heptachlor on several occasions and in 1991 established an ADI of 0.1 µg/kg of body weight on the basis of a NOAEL of 0.025 mg/kg of body weight per day from two studies in the dog, incorporating an uncertainty factor of 200 (100 for inter- and intraspecies variation and 2 for the inadequacy of the database). With an allocation of 1% of the ADI to drinking-water, because the main source of exposure seems to be food, the guideline value is 0.03 µg/litre.

### Hexachlorobenzene

Hexachlorobenzene (HCB) has been used as a selective fungicide, but its use is now uncommon. It is a by-product of several chemical processes and an impurity in some pesticides. HCB is strongly adsorbed by soil and sediments and has a half-life measured in years. It is a ubiquitous contaminant and is readily lost to the atmosphere. It is resistant to degradation and has a high accumulation potential, accumulating in the tissues of aquatic and terrestrial organisms.

Food is considered to be the major source of exposure to HCB. Atmospheric contamination may also contribute to the intake of HCB by humans. HCB has not been found in drinking-water.

In 1987, IARC reviewed data on the carcinogenicity of HCB and assigned it to Group 2B. Because HCB has been shown to induce tumours in three animal species and at a variety of sites, a linearized low-dose extrapolation model was used to calculate the guideline value. On the basis of liver tumours observed in female rats in a 2-year dietary study and applying the linearized multistage model, a guideline value in drinking-water of 1 µg/litre, corresponding to an excess lifetime cancer risk of $10^{-5}$, was calculated.

### Isoproturon

Isoproturon is a selective, systemic herbicide used in the control of annual grasses and broad-leaved weeds in cereals. It can be photodegraded, hydrolysed, and biodegraded and persists from days to weeks. It is mobile in soil and has been detected in surface and ground water. There is evidence that exposure to this compound through food is low.

Isoproturon is of low acute toxicity and low to moderate toxicity following short- and long-term exposures. It does not possess significant genotoxic activity, but it causes marked enzyme induction and liver enlargement. Isoproturon caused an increase in hepatocellular tumours in male and female rats, but this was apparent only at doses that also caused liver toxicity. Isoproturon appears to be a tumour promoter rather than a complete carcinogen.

On the basis of this evaluation, it is appropriate to derive a guideline by calculating a TDI using an uncertainty factor. The NOAELs in a 90-day study in dogs and a 2-year feeding study in rats were approximately 3 mg/kg of body weight per day. A TDI of 3 µg/kg of body weight can be calculated by applying an uncertainty factor of 1000 (100 for inter- and intraspecies variation and 10 because there is evidence of non-genotoxic carcinogenicity in rats). A guideline value of 9 µg/litre was calculated by allocating 10% of the TDI to drinking-water.

### Lindane

Lindane ($\gamma$-hexachlorocyclohexane, $\gamma$-HCH) is an insecticide that has been used for a very long time. Apart from agricultural uses on plants and animals, it is also used in public health and as a wood preservative.

Lindane is a persistent compound with a relatively low affinity for water and a low mobility in soil; it slowly volatilizes into the atmosphere. It is a ubiquitous environmental contaminant, and has been detected in water. Exposure of humans occurs mainly via food, but this is decreasing.

Lindane causes liver tumours in mice given very high doses, but there is evidence that this is a result of tumour promotion. In 1987, IARC classified lindane in Group 2B. Moreover, in 1989, after reviewing all available *in vitro* and *in vivo* short-term tests, JMPR concluded that there was no evidence of genotoxicity and established an ADI of 8 µg/kg of body weight based on liver and kidney toxicity observed in a short-term study in the rat.

On the basis of the same study, but using a compound intake estimate considered to be more appropriate in the light of additional data, a TDI of 5 µg/kg of body weight was derived from a NOAEL of 0.5 mg/kg of body weight per day by applying an uncertainty factor of 100 (for inter- and intraspecies variation). It was not considered necessary to include an additional uncertainty factor to allow for the tumour-promoting potential in view of the substantial database and numerous international evaluations of this compound supporting the TDI.

Although exposure from food is decreasing, there may be substantial exposure from its use in public health and as a wood preservative. Therefore, only 1% of the TDI was allocated to drinking-water. The guideline value is thus 2 µg/litre (rounded figure).

### MCPA

MCPA is a chlorophenoxy post-emergence herbicide that is very soluble, is highly mobile, and can leach from the soil. It is metabolized by bacteria and can be photochemically degraded. MCPA has only limited persistence and has not been frequently detected in drinking-water.

There are only limited and inconclusive data on the genotoxicity of MCPA. IARC evaluated MCPA in 1983 and concluded that the available data on humans and experimental animals were inadequate for an evaluation of carcinogenicity. Further evaluations by IARC on chlorophenoxy herbicides in 1986 and 1987 concluded that evidence for their carcinogenicity was limited in humans and inadequate in animals (Group 2B). Recent carcinogenicity studies on rats and mice did not indicate that MCPA was carcinogenic. No adequate epidemiological data on exposure to MCPA alone are available.

Long-term toxicity studies in rats and mice and a 1-year feeding study in dogs are available. The NOAEL was 0.15 mg/kg of body weight per day in the study in dogs, based on renal and liver toxicity observed at higher doses levels. A TDI of 0.5 $\mu$g/kg of body weight was established based on the NOAEL from the 1-year study and an uncertainty factor of 300 (100 for intra- and interspecies variation and 3 for the inadequacy of the database). An allocation of 10% of the TDI to drinking-water results in a guideline value of 2 $\mu$g/litre (rounded figure).

Methoxychlor

Methoxychlor is an insecticide used on vegetables, fruit, trees, fodder, and farm animals. It is poorly soluble in water and highly immobile in most agricultural soils. Under normal conditions of use, methoxychlor seems not to be of environmental concern. However, it has been detected occasionally in drinking-water. Daily intake from food and air is expected to be below 1 $\mu$g per person.

Environmental metabolites are formed preferentially under anaerobic rather than aerobic conditions and include mainly the dechlorinated and demethylated products. There is some potential for the accumulation of the parent compound and its metabolites in surface water sediments.

The genotoxic potential of methoxychlor appears to be negligible. In 1979, IARC assigned methoxychlor to Group 3. Subsequent data suggest a carcinogenic potential of methoxychlor for liver and testes in mice. This may be due to the hormonal activity of proestrogenic mammalian metabolites of methoxychlor and may therefore have a threshold. The study, however, was inadequate because only one dose was used and because this dose may have been above the maximum tolerated dose.

The database for studies on long-term, short-term, and reproductive toxicity is inadequate. A teratology study in rabbits reported a systemic NOAEL of 5 mg/kg of body weight per day, which is lower than the LOAELs and NOAELs from other studies. This NOAEL was therefore selected for use in the derivation of a TDI.

The application of an uncertainty factor of 1000 (100 for inter- and intraspecies differences and 10 for concern for threshold carcinogenicity and the limited database) leads to a TDI of 5 $\mu$g/kg of body weight. Allocation of 10% of the TDI to drinking-water results in a guideline value of 20 $\mu$g/litre (rounded figure).

### Metolachlor

Metolachlor is a selective pre-emergence herbicide used on a number of crops. It can be lost from the soil through biodegradation, photodegradation, and volatilization. It is fairly mobile and under certain conditions can contaminate ground water, but it is mostly found in surface water.

There is no evidence from available studies that metolachlor is carcinogenic in mice. In rats, an increase in liver tumours in females as well as a few nasal tumours in males have been observed. Metolachlor is not genotoxic.

Toxicity data were available from long-term studies in rodents and from a 1-year study in dogs. An apparent decrease in kidney weight was observed at the two highest dose levels in the 1-year dog study, giving a NOAEL of 3.5 mg/kg of body weight per day. Applying an uncertainty factor of 1000 to this NOAEL (100 for inter- and intraspecies variation and 10 because of some concern regarding carcinogenicity), a TDI of 3.5 µg/kg of body weight was derived. A 10% allocation of the TDI to drinking-water results in a guideline value of 10 µg/litre (rounded figure).

### Molinate

Molinate is a herbicide used to control broad-leaved and grassy weeds in rice. The available data suggest that ground water pollution by molinate is restricted to some rice-growing regions. Data on the occurrence of molinate in the environment are limited but indicate that concentrations in water rarely exceed 1 µg/litre. Molinate is of low persistence in water and soil, with a half-life of about 5 days.

On the basis of the limited information available, molinate does not seem to be carcinogenic or mutagenic in animals. Evidence suggests that impairment of the reproductive performance of the male rat represents the most sensitive indicator of molinate exposure. However, epidemiological data based on the examination of workers involved in molinate production do not indicate any effect on human fertility.

The NOAEL for reproductive toxicity in the rat was 0.2 mg/kg of body weight per day, and this value was chosen as the basis for calculating a TDI for molinate. Using an uncertainty factor of 100 (for inter- and intraspecies variation), a TDI of 2 µg/kg of body weight was derived. An allocation of 10% of the TDI to drinking-water results in a guideline value of 6 µg/litre.

### Pendimethalin

Pendimethalin is a pre-emergence herbicide that is fairly immobile and persistent in soil. It is lost through photodegradation, biodegradation, and volatilization. The leaching potential of pendimethalin appears to be very low, but little is known about its more polar degradation products. It has rarely been found in drinking-water in the limited studies available.

On the basis of available data, pendimethalin does not appear to have significant mutagenic activity. Long-term studies in mice and rats have not provided evidence of carcinogenicity; however, these studies have some important limitations.

In a long-term rat feeding study, evidence of slight liver toxicity was noted even at the lowest dose tested; a NOAEL for this finding was not established. The LOAEL was 5 mg/kg of body weight per day. Applying an uncertainty factor of 1000 (100 for intra- and interspecies variation and 10 for the use of a LOAEL instead of a NOAEL and for limitations in the database), a TDI of 5 µg/kg of body weight was calculated. An allocation of 10% of the TDI to drinking-water results in a guideline value of 20 µg/litre (rounded figure).

Pentachlorophenol
Pentachlorophenol (PCP) is used mainly as a wood preservative. Elevated PCP concentrations can be found in ground water and surface water within wood treatment areas. The general population is exposed to PCP through the ingestion of drinking-water and food, as well as through exposure to treated items (e.g., textiles, leather and paper products) and, above all, inhalation of indoor air contaminated with PCP.

Unpurified technical PCP contains several microcontaminants, particularly polychlorinated dibenzo-*p*-dioxins (PCDDs) and dibenzofurans (PCDFs), of which hexachlorodibenzo-*p*-dioxin is the most relevant congener toxicologically.

In short- and long-term animal studies, exposure to relatively high PCP concentrations has been shown to reduce growth rates and serum thyroid hormone levels and to increase liver weights and liver enzyme activity. Exposure to much lower concentrations of technical PCP formulations has been shown to decrease growth rates, increase weights of liver, lungs, kidneys and adrenal glands, increase liver enzyme activity, interfere with porphyrin metabolism and renal function, and alter haematological and biochemical parameters. Microcontaminants appear to be the principal active moieties in the nonacute toxicity of commercial PCP.

PCP has been shown to be fetotoxic, delaying the development of rat embryos and reducing litter size, neonatal body weight and survival, and weanling growth. The NOAEL for technical PCP was a maternal dose of 5 mg/kg of body weight per day during organogenesis. PCP is not considered to be teratogenic, although birth defects arose as an indirect result of maternal hyperthermia in one study. The NOAEL in rat reproduction studies was 3 mg/kg of body weight per day. This value is close to the NOAEL in the fetotoxicity study, but there are no corroborating studies in other mammalian species.

PCP has been shown to be immunotoxic in several animal species. At least part of this effect is caused by PCP itself. Neurotoxic effects have also been reported, but the possibility that these are due to microcontaminants has not been excluded.

## 3. CHEMICAL ASPECTS

Pure PCP has not been found to be highly mutagenic. The presence of at least one carcinogenic microcontaminant (hexachlorodibenzo-*p*-dioxin) suggests that the potential for technical PCP to cause cancer in laboratory animals cannot be completely ruled out.

The NOAEL of 3 mg/kg of body weight per day was used to calculate the guideline value. An uncertainty factor of 1000 (100 for intra- and interspecies variation and 10 for potential carcinogenicity of technical PCP) was applied to derive a TDI of 3 µg/kg of body weight. An allocation of 10% of the TDI to drinking-water gives a guideline value of 9 µg/litre. This guideline value is considered provisional, because PCP was evaluated only at the final Task Group meeting (see Annex 1), on the basis of Environmental Health Criteria No. 71.[1]

### Permethrin

Permethrin is a synthetic pyrethroid insecticide that is widely used in crop protection and public health. It is used in water reservoirs for mosquito larvae control and for control of infestation of water mains by aquatic invertebrates.

Permethrin has a marked affinity for soil and sediment and a low affinity for water, and it is not likely to be lost to the atmosphere. It can be photodegraded and biodegraded, and it persists for periods ranging from days to weeks.

Permethrin does not accumulate in mammals because of its rapid metabolism. Exposure to permethrin in food and through household and public health use is likely to be high.

Permethrin is of low mammalian toxicity. It is usually used as a mixture of the *cis* and *trans* isomers; the *cis*-isomer, which is the active component, is more toxic than the *trans*-isomer.

Permethrin is not genotoxic. Although there was a slightly increased incidence of benign lung tumours in male mice in one study, this was only at the highest dose and was not considered to indicate any significant carcinogenic potential for permethrin. IARC has classified permethrin in Group 3.

A TDI approach can be used to calculate a guideline value. In 1987, JMPR recommended an ADI for 2:3 and 1:3 *cis*:*trans*-permethrin of 0.05 mg/kg of body weight based on the application of an uncertainty factor of 100 to a NOAEL for liver toxicity equivalent to 5 mg/kg of body weight per day.

Because there is significant exposure to permethrin from the environment, only 1% of the ADI is allocated to drinking-water. Therefore, the guideline value is 20 µg/litre (rounded figure). However, if permethrin is to be used as a larvicide for the control of mosquitos and other insects of health significance in drinking-water sources, the share of the ADI allocated to drinking-water may be increased.

---

[1] *Pentachlorophenol*. Geneva, World Health Organization, 1987 (Environmental Health Criteria, No. 71). An evaluation document on PCP has not been prepared for Volume 2 of the *Guidelines*.

### Propanil

Propanil is a contact post-emergence herbicide used to control broad-leaved and grassy weeds, mainly in rice. It is a mobile compound with affinity for the water compartment. Propanil is not, however, persistent, being easily transformed under natural conditions to several metabolites. Two of these metabolites, 3,4-dichloroaniline and 3,3',4,4'-tetrachloroazobenzene (TCAB), are more toxic and more persistent than the parent compound. Although used in a number of countries, propanil has only occasionally been detected in ground water.

Propanil is considered not to be genotoxic. However, at least one of its environmental metabolites (TCAB) is genotoxic. Data from a limited study in rats do not provide evidence of carcinogenicity.

Long-term exposure to propanil results in red blood cell toxicity. A TDI of 5 µg/kg of body weight was established, based on the NOAEL of 5 mg/kg of body weight per day from a 3-month rat feeding study and applying an uncertainty factor of 1000 (100 for intra- and interspecies variation and 10 for the short duration of the study and limitations of the database).

Based on an allocation of 10% of the TDI to drinking-water, the guideline value is 20 µg/litre (rounded figure). In applying this guideline, authorities should consider the possible presence of more toxic metabolites in water.

### Pyridate

Pyridate is a contact herbicide used in cereals, maize, rice, and other crops. It has very low water solubility and relatively low mobility. It is not persistent and is rapidly hydrolysed, photodegraded, and biodegraded. Its primary environmental metabolite is also not persistent but is more mobile. Under favourable conditions, the environmental half-life is of the order of a few days. This compound is only rarely found in drinking-water.

The available evidence indicates that pyridate is not genotoxic. Pyridate has been tested in long-term feeding studies in rats and mice; no evidence of carcinogenicity was noted in either species.

The NOAEL of 3.5 mg/kg of body weight per day in a 2-year rat study is based upon increased kidney weight. A TDI of 35 µg/kg of body weight was calculated by applying an uncertainty factor of 100 (for intra- and interspecies variation) to this NOAEL. An allocation of 10% of the TDI to drinking-water gives a guideline value of 100 µg/litre (rounded figure).

### Simazine

Simazine is a pre-emergence herbicide used on a number of crops as well as in non-crop areas. It is fairly resistant to physical and chemical dissipation processes in the soil. Its persistence and mobility are such that it has been frequently detected in ground and surface waters at concentrations of up to a few micrograms per litre.

Simazine does not appear to be genotoxic in mammalian systems. Recent studies have shown an increase in mammary tumours in the female rat but no effects in the mouse. IARC has classified simazine in Group 3.

Based on a study in the rat, a NOAEL of 0.52 mg/kg of body weight per day has been established for carcinogenicity and long-term toxicity. By applying an uncertainty factor of 1000 (100 for intra- and interspecies variation and 10 for possible carcinogenicity), a TDI of 0.52 µg/kg of body weight was derived. An allocation of 10% of the TDI to drinking-water gives a guideline value of 2 µg/litre (rounded figure).

## Trifluralin

Trifluralin is a pre-emergence herbicide used in a number of crops. It has low water solubility and a high affinity for soil. However, biodegradation and photodegradation processes may give rise to polar metabolites that may contaminate drinking-water sources. Although this compound is used in many countries, relatively few data are available concerning contamination of drinking-water. Trifluralin was not detected in the small number of samples analysed.

Trifluralin of high purity does not possess mutagenic properties. Technical trifluralin of low purity may contain nitroso contaminants and has been found to be mutagenic. No evidence of carcinogenicity was demonstrated in a number of long-term toxicity/carcinogenicity studies with pure (99%) test material. IARC recently evaluated technical-grade trifluralin and assigned it to Group 3.

A NOAEL of 0.75 mg/kg of body weight per day was selected based on a 1-year feeding study in dogs. This species is the most sensitive for the mild hepatic effects on which the NOAEL was based. Using this NOAEL and an uncertainty factor of 100 (for intra- and interspecies variation), a TDI of 7.5 µg/kg of body weight was derived. A guideline value of 20 µg/litre (rounded figure) is recommended based on an allocation of 10% of the TDI to drinking-water.

Authorities should note that some impure technical grades of trifluralin could contain potent carcinogenic compounds and therefore should not be used.

## Chlorophenoxy herbicides (excluding 2,4-D and MCPA)

The chlorophenoxy herbicides considered here are 2,4-DB, dichlorprop, fenoprop, MCPB, mecoprop, and 2,4,5-T. The half-lives for degradation of these compounds in the environment are of the order of several days. Limited monitoring data indicate that these herbicides are not frequently found in drinking-water; when detected, their concentrations are usually no greater than a few micrograms per litre. These chlorophenoxy herbicides are not often found in food.

Chlorophenoxy herbicides, as a group, have been classified in Group 2B by IARC. However, the available data from studies in exposed populations and animals do not permit assessment of the carcinogenic potential to humans of any specific chlorophenoxy herbicide. Therefore, drinking-water guidelines for these compounds are based on a threshold approach for other toxic effects.

*2,4-DB*

In a 2-year study in rats, the NOAEL for effects on body and organ weights, blood chemistry, and haematological parameters was determined to be 3 mg/kg of body weight per day. A TDI of 30 µg/kg of body weight was derived using an uncertainty factor of 100 (for intra- and interspecies variation). With the allocation of 10% of the TDI to drinking-water, the guideline value is 90 µg/litre.

*Dichlorprop*

Based on a 2-year study in rats, the NOAEL for renal toxicity is 3.64 mg/kg of body weight per day. The TDI for dichlorprop was calculated to be 36.4 µg/kg of body weight by applying an uncertainty factor of 100 (for intra- and interspecies variation) to this NOAEL. With the allocation of 10% of the TDI to drinking-water, the guideline value is 100 µg/litre (rounded figure).

*Fenoprop*

A NOAEL of 0.9 mg/kg of body weight per day for adverse effects on the liver was reported in a study in which beagle dogs were administered fenoprop in the diet for 2 years. A TDI of 3 µg/kg of body weight was derived using an uncertainty factor of 300 (100 for intra- and interspecies variation and 3 for limitations of the database). With the allocation of 10% of the TDI to drinking-water, the guideline value for fenoprop is 9 µg/litre.

*MCPB*

Currently available toxicological data are insufficient to be used as the basis for a guideline value for MCPB in drinking-water.

*Mecoprop*

A NOAEL of 1 mg/kg of body weight per day for effects on kidney weight in 1- and 2-year studies in rats was used with an uncertainty factor of 300 (100 for intra- and interspecies variation and 3 for limitations of the database) to derive a TDI of 3.33 µg/kg of body weight. With the allocation of 10% of the TDI to drinking-water, the guideline value for mecoprop is 10 µg/litre (rounded figure).

*2,4,5-T*

The NOAEL for reduced body weight gain, increased liver and kidney weights, and renal toxicity in a 2-year study in rats was 3 mg/kg of body weight per day. A TDI of 3 µg/kg of body weight was derived using an uncertainty factor of 1000 (100 for intra- and interspecies variation and 10 for the suggested association between 2,4,5-T and soft tissue sarcoma and non-Hodgkin lymphoma in epidemiological studies). With the allocation of 10% of the TDI to drinking-water, the guideline value for 2,4,5-T is 9 µg/litre.

### 3.6.4 Disinfectants and disinfectant by-products

Disinfection is unquestionably the most important step in the treatment of water for public supply. The destruction of microbiological pathogens is essential and almost invariably involves the use of reactive chemical agents such as chlorine, which are not only powerful biocides but also capable of reacting with other water constituents to form new compounds with potentially harmful long-term health effects. Thus, an overall assessment of the impact of disinfection on public health must consider not only the microbiological quality of the treated water, but also the toxicity of the disinfectants and their reaction products.

The paramount importance of microbiological quality requires some flexibility in the derivation of guideline values for these substances. Fortunately this is possible because of the substantial margin of safety incorporated into these values. Guideline values for carcinogenic disinfectant by-products are presented here for an excess lifetime cancer risk of $10^{-5}$. The conditions specified for disinfection vary not only according to water composition and temperature but also with available technology and socioeconomic factors in different parts of the world. Where local circumstances require that a choice must be made between meeting either microbiological guidelines or guidelines for disinfectants or disinfectant by-products, the microbiological quality must always take precedence, and where necessary, a chemical guideline value can be adopted corresponding to a higher level of risk. Efficient disinfection must *never* be compromised.

Although not addressed with respect to the individual parameters presented below, it is noted that, in a number of epidemiological studies, positive associations between the ingestion of chlorinated drinking-water and mortality rates from cancer, particularly of the bladder, have been reported. The degree of evidence for this association is considered inadequate by IARC.

The level of disinfection by-products can be reduced by optimizing the treatment process (see section 6.3). Removal of organic substances prior to disinfection reduces the formation of potentially harmful by-products.

The following guidance is provided to help authorities decide which guideline values may be of greater or lesser importance for setting national standards: guideline values for chemicals of greater importance generally include those for chloramines and chlorine (when used as disinfectants); followed by those for bromoform, dibromochloromethane, bromodichloromethane, chloroform, and chloral hydrate; and chlorite, bromate, dichloroacetic acid, and trichloroacetic acid (provisional guideline values have been established for this last group). Guideline values for chemicals of lesser importance generally include those for 2,4,6-trichlorophenol, formaldehyde, dichloroacetonitrile, dibromoacetonitrile, trichloroacetonitrile, and cyanogen chloride. Although given less importance, it may be appropriate to measure their levels at least once. It should also be noted that a number of non-volatile, poorly characterized by-products may be formed as well, including those derived from humic substances. These recommendations are general, and local monitoring and surveillance capabilities must be considered in the setting of national standards.

## Disinfectants

### Chloramines

Monochloramine is present in drinking-water as a disinfectant and as a by-product of chlorination. Drinking-water is the major source of exposure to chloramines.

Adverse health effects have not been observed following short-term exposure of humans to concentrations of up to 24 mg/litre. In addition, in short- and long-term studies in laboratory animals exposed to monochloramine, no specific, clearly adverse treatment-related effects have been observed.

In a bioassay in two species, the incidence of mononuclear-cell leukaemias in female F344 rats was increased in comparison with concurrent controls but was within the range of that observed in historical controls. No other increases in tumour incidence were observed. Although monochloramine has been shown to be mutagenic in some *in vitro* studies, it has not been found to be genotoxic *in vivo*.

The guideline value for monochloramine is based on a TDI of 94 µg/kg of body weight, calculated from a NOAEL of 9.4 mg/kg of body weight per day (the highest dose administered to males in the rat study) and incorporating an uncertainty factor of 100 (for intra- and interspecies variation). An additional uncertainty factor for possible carcinogenicity was not applied because equivocal cancer effects reported in the same study in only one species and in only one sex were within the range observed in historical controls. With an allocation of 100% of the TDI to drinking-water, the guideline value is 3 mg/litre (rounded figure).

Available data are insufficient for the establishment of guideline values for dichloramine and trichloramine. The odour thresholds for dichloramine and trichloramine are much lower than that for monochloramine.

### Chlorine

Chlorine is produced in large amounts and widely used both industrially and domestically as a disinfectant and bleach. In particular, it is widely used in the disinfection of swimming-pools and is the most commonly used disinfectant and oxidant in drinking-water treatment. In water, chlorine reacts to form hypochlorous acid and hypochlorites.

In humans and animals exposed to chlorine in drinking-water, no specific adverse treatment-related effects have been observed. IARC has classified hypochlorite in Group 3.

The guideline value for free chlorine in drinking-water is based on a TDI of 150 µg/kg of body weight, derived from a NOAEL for the absence of toxicity in rodents ingesting 15 mg of chlorine per kg of body weight per day in drinking-water for 2 years and incorporating an uncertainty factor of 100 (for intra- and interspecies variation). With an allocation of 100% of the TDI to drinking-water, the guideline value is 5 mg/litre (rounded figure). It should be noted, however, that this value is conservative, as no adverse effect level was identified in this study. Most individuals are able to taste chlorine at the guideline value (see page 129).

### Chlorine dioxide

Chlorine dioxide is a strong oxidizing agent that is added to water as a disinfectant and to control taste and odour. Chlorine dioxide rapidly decomposes into chlorite, chloride, and chlorate.

Chlorine dioxide has been shown to impair neurobehavioural and neurological development in rats exposed perinatally. Significant depression of thyroid hormones has also been observed in rats and monkeys exposed to chlorine dioxide in drinking-water studies.

A guideline value has not been established for chlorine dioxide because of its rapid breakdown and because the chlorite provisional guideline value (see page 96) is adequately protective for potential toxicity from chlorine dioxide. The taste and odour threshold for this compound is 0.4 mg/litre.

### Iodine

Iodine occurs naturally in water in the form of iodide. Traces of iodine are produced by oxidation of iodide during water treatment. Iodine is occasionally used for water disinfection in the field or in emergency situations.

Iodine is an essential element for the synthesis of thyroid hormones. Estimates of the dietary requirement for adult humans range from 80 to 150 µg/day; in many parts of the world, there are dietary deficiencies in iodine. In 1988, JECFA set a PMTDI for iodine of 1 mg/day (17 µg/kg of body weight per day) from all sources, based primarily on data on the effects of iodide. However, recent data from studies in rats indicate that the effects of iodine in drinking-water on thyroid hormone concentrations in the blood differ from those of iodide.

Available data therefore suggest that derivation of a guideline value for iodine on the basis of information on the effects of iodide is inappropriate, and there are few relevant data on the effects of iodine. Because iodine is not recommended for long-term disinfection, lifetime exposure to iodine concentrations such as might occur from water disinfection is unlikely. For these reasons, a guideline value for iodine has not been established at this time.

## Disinfectant by-products

### Bromate

Bromate can be formed by the oxidation of bromide ions during ozonation and possibly by other oxidants in water treatment. Limited data indicate that concentrations in drinking-water are generally less than 90 $\mu$g/litre.

Bromate has been found to induce a very high incidence of kidney tumours in male and female rats and peritoneal mesotheliomas in male rats. Bromate is mutagenic *in vitro* and *in vivo*. JECFA evaluated bromate and recommended that there should be no residues in food when bromate is used in food processing.

IARC has classified bromate in Group 2B. To estimate cancer risks, the linearized multistage model was applied to the incidence of renal tumours in male rats given potassium bromate in drinking-water, although it was noted that if the mechanism of tumour induction is determined to be oxidative damage in the kidney, the application of the low-dose cancer risk model may not be appropriate. The concentration in drinking-water associated with an excess lifetime cancer risk of $10^{-5}$ is 3 $\mu$g/litre. Because of limitations in available analytical and treatment methods, a provisional guideline value of 25 $\mu$g/litre is recommended. This value is associated with an excess lifetime cancer risk of $7 \times 10^{-5}$.

### Chlorate

In addition to being a decomposition product of chlorine dioxide, chlorate also occurs as a result of the use of hypochlorite for disinfection. Available data on the effects of chlorate in humans and experimental animals are considered insufficient to permit development of a guideline value. Data on accidental poisonings indicate that the lethal dose to humans is about 230 mg/kg of body weight per day. This is of the same order of magnitude as the NOAELs identified from studies in rats and dogs. Although no effects were observed in an 84-day clinical study in a small number of human volunteers ingesting 36 $\mu$g/kg of body weight per day, a guideline value was not derived on the basis of these results because no adverse effect level was determined.

Further research is needed to characterize the nonlethal effects of chlorate. Until data become available, it may be prudent to try to minimize chlorate levels. However, adequate disinfection should not be compromised.

### Chlorite

Chlorite affects red blood cells, resulting in methaemoglobin formation in cats and monkeys. IARC has classified chlorite in Group 3.

The TDI for chlorite is 10 µg/kg of body weight, based on the NOAEL of 1 mg/kg of body weight per day for decreased glutathione levels in a 90-day study in rats and incorporating an uncertainty factor of 100 (for intra- and interspecies variation). Owing to the acute nature of the response and the existence of a 2-year rat study, an additional uncertainty factor of 10 was not incorporated to account for the short duration of the key study. The TDI derived in this manner is consistent with the NOAEL (36 µg/kg of body weight per day) in a 12-week clinical study in a small number of human volunteers.

Allocating 80% of the TDI to drinking-water gives a provisional guideline value of 200 µg/litre (rounded figure). This guideline value is designated as provisional because use of chlorine dioxide as a disinfectant may result in the chlorite guideline value being exceeded, and difficulties in meeting the guideline value must never be a reason for compromising adequate disinfection.

### Chlorophenols

Chlorophenols are present in drinking-water as a result of chlorination of phenols, as by-products of the reaction of hypochlorite with phenolic acids, as biocides, or as degradation products of phenoxy herbicides. Those most likely to occur in drinking-water as by-products of chlorination are 2-chlorophenol (2-CP), 2,4-dichlorophenol (2,4-DCP), and 2,4,6-trichlorophenol (2,4,6-TCP).

Concentrations of chlorophenols in drinking-water are usually less than 1 µg/litre. The taste thresholds for chlorophenols in drinking-water are low (see page 130).

#### 2-Chlorophenol

Data on the toxicity of 2-CP are limited. Therefore, no health-based guideline value has been derived.

#### 2,4-Dichlorophenol

Data on the toxicity of 2,4-DCP are limited. Therefore, no health-based guideline value has been derived.

#### 2,4,6-Trichlorophenol

2,4,6-TCP has been reported to induce lymphomas and leukaemias in male rats and hepatic tumours in male and female mice. The compound has not been shown to be mutagenic in the Ames test but has shown weak mutagenic activity in other *in vitro* and *in vivo* studies. IARC has classified 2,4,6-TCP in Group 2B.

A guideline value can be derived for 2,4,6-TCP by applying the linearized multistage model to leukaemias in male rats observed in a 2-year feeding study. The hepatic tumours found in this study were not used for risk estimation, because of the possible role of contaminants in their induction. The concentration in drinking-water associated with a $10^{-5}$ excess lifetime cancer risk is 200 µg/litre. This concentration exceeds the lowest reported taste threshold for 2,4,6-TCP (see page 130).

Formaldehyde

Formaldehyde occurs in industrial effluents and is emitted into air from plastic materials and resin glues. Formaldehyde in drinking-water results primarily from the oxidation of natural organic matter during ozonation and chlorination. It is also found in drinking-water as a result of release from polyacetal plastic fittings. Concentrations of up to 30 µg/litre have been found in ozonated drinking-water.

Formaldehyde has been shown to be carcinogenic in rats and mice by inhalation at doses that caused irritation of the nasal epithelium. Ingestion of formaldehyde in drinking-water for 2 years caused stomach irritation in rats, and papillomas of the stomach associated with severe irritation were observed in one study.

On the basis of studies in which humans and experimental animals were exposed by inhalation, IARC has classified formaldehyde in Group 2A. The weight of the evidence indicates that formaldehyde is not carcinogenic by the oral route. A guideline value has been derived, therefore, on the basis of a TDI. A TDI of 150 µg/kg of body weight was calculated based on the NOAEL of 15 mg/kg of body weight per day in a 2-year study in rats, incorporating an uncertainty factor of 100 (for intra- and interspecies variation). No account was taken of potential carcinogenicity from the inhalation of formaldehyde from various indoor water uses, such as showering (see section 3.3). With an allocation of 20% of the TDI to drinking-water, the guideline value is 900 µg/litre.

MX

MX, or 3-chloro-4-dichloromethyl-5-hydroxy-2(5H)-furanone, is formed by the reaction of chlorine with complex organic matter in water. It has been identified in chlorinated effluents of pulp mills and drinking-water in Finland, the United Kingdom and the United States of America at concentrations of up to 67 ng/litre.

There are very limited data on the toxicity of MX. $^{14}$C-labelled MX is rapidly adsorbed, and most of the radioactivity is excreted in the urine within 24–48 hours. It is unlikely to be absorbed as the parent compound because of its high reactivity. MX is an extremely potent mutagen in some strains of *Salmonella typhimurium*, but the addition of liver extract dramatically reduces the response. It is only weakly active or non-active in short-term tests for genotoxicity *in vivo*.

Available data are inadequate to permit a guideline value for MX to be established.

## 3. CHEMICAL ASPECTS

### Trihalomethanes

Trihalomethanes are halogen-substituted single-carbon compounds with the general formula $CHX_3$, where X may be fluorine, chlorine, bromine, or iodine, or a combination thereof. With respect to drinking-water contamination, only four members of the group are important: bromoform, dibromochloromethane (DBCM), bromodichloromethane (BDCM), and chloroform. The most commonly occurring constituent is chloroform.

Trihalomethanes occur in drinking-water principally as products of the reaction of chlorine with naturally occurring organic materials and with bromide, which may also be present in the water.

This group of chemicals may act as an indicator for the presence of other chlorination by-products. Control of these four trihalomethanes should help to reduce levels of other uncharacterized chlorination by-products.

Because these four compounds usually occur together, it has been the practice to consider total trihalomethanes as a group, and a number of countries have set guidelines or standards on this basis. In the first edition of the *Guidelines for drinking-water quality*, a guideline value was established for chloroform only: few data existed for the remaining trihalomethanes, and, for most water supplies, chloroform was the most commonly encountered member of the group. In this edition, no guideline value has been set for total trihalomethanes; however, guideline values have been established separately for all four trihalomethanes. For authorities wishing to establish a total trihalomethane standard to account for additive toxicity, the following fractionation approach could be taken:

$$\frac{C_{bromoform}}{GV_{bromoform}} + \frac{C_{DBCM}}{GV_{DBCM}} + \frac{C_{BDCM}}{GV_{BDCM}} + \frac{C_{chloroform}}{GV_{chloroform}} \leqslant 1$$

where $C$ = concentration and $GV$ = guideline value.

Authorities wishing to use a guideline value for total trihalomethanes should not simply add up the guideline values for the individual compounds in order to arrive at a standard, because the four compounds are basically similar in toxicological action.

In controlling trihalomethanes, a multistep treatment system should be used to reduce organic trihalomethane precursors, and primary consideration should be given to ensuring that disinfection is never compromised.

### Bromoform

Bromoform is readily absorbed from the gastrointestinal tract. In experimental animals, long-term exposure to high doses causes damage to the liver and kidney. In one bioassay, bromoform induced a small increase in relatively rare tumours of the large intestine in rats of both sexes but did not induce tumours in mice.

Data from a variety of assays on the genotoxicity of bromoform are equivocal. IARC has classified bromoform in Group 3.

A TDI was derived on the basis of a NOAEL of 25 mg/kg of body weight per day for the absence of histopathological lesions in the liver in a well-conducted and well-documented 90-day study in rats. This NOAEL is supported by the results of two long-term studies. The TDI is 17.9 µg/kg of body weight, correcting for exposure on 5 days per week and using an uncertainty factor of 1000 (100 for intra- and interspecies variation and 10 for possible carcinogenicity and the short duration of the study). With an allocation of 20% of the TDI to drinking-water, the guideline value is 100 µg/litre (rounded figure).

*Dibromochloromethane*

Dibromochloromethane is well absorbed from the gastrointestinal tract. In experimental animals, long-term exposure to high doses causes damage to the liver and kidney. In one bioassay, dibromochloromethane induced hepatic tumours in female and possibly in male mice but not in rats. The genotoxicity of dibromochloromethane has been studied in a number of assays, but the available data are considered inconclusive. IARC has classified dibromochloromethane in Group 3.

A TDI was derived on the basis of a NOAEL of 30 mg/kg of body weight per day for the absence of histopathological effects in the liver in a well-conducted and well-documented 90-day study in rats. This NOAEL is supported by the results of long-term studies. The TDI is 21.4 µg/kg of body weight, correcting for exposure on 5 days per week and using an uncertainty factor of 1000 (100 for intra- and interspecies variation and 10 for the short duration of the study). An additional uncertainty factor for potential carcinogenicity was not applied because of the questions regarding mice liver tumours from corn oil vehicles and inconclusive evidence of genotoxicity. With an allocation of 20% of the TDI to drinking-water, the guideline value is 100 µg/litre (rounded figure).

*Bromodichloromethane*

Bromodichloromethane is readily absorbed from the gastrointestinal tract. In experimental animals, long-term exposure to high doses causes damage to the liver and kidney. In one bioassay, bromodichloromethane induced renal adenomas and adenocarcinomas in both sexes of rats and male mice, rare tumours of the large intestine (adenomatous polyps and adenocarcinomas) in both sexes of rats, and hepatocellular adenomas and adenocarcinomas in female mice. Bromodichloromethane has given both positive and negative results in a variety of *in vitro* and *in vivo* genotoxicity assays. IARC has classified bromodichloromethane in Group 2B.

Cancer risks have been estimated on the basis of increases in incidence of kidney tumours in male mice observed in the bioassay described above, as these tumours yield the most protective value. Hepatic tumours in female mice were not considered owing to the possible role of the corn oil vehicle in induction of these tumours, although the estimated risks are within the same range. Using the linearized multistage model, the concentration in drinking-water associated with an excess lifetime cancer risk of $10^{-5}$ is 60 µg/litre. This guideline value is supported by a recently published feeding study in rats that was not available for full evaluation.

*Chloroform*
Chloroform concentrations in drinking-water can sometimes range up to several hundred micrograms per litre. Concentrations in ambient air are usually low, and chloroform has been detected in some foods at levels usually in the range of 1–30 µg/kg.

Chloroform is absorbed following oral, inhalation, and dermal exposure, and several reactive metabolic intermediates can be produced, the extent of which varies with species and sex. Long-term exposure to dose levels in excess of 15 mg/kg of body weight per day can cause changes in the kidney, liver, and thyroid.

IARC has classified chloroform in Group 2B. In long-term studies, chloroform has been shown to induce hepatocellular carcinomas in mice when administered by gavage in oil-based vehicles but not in drinking-water; it has been reported to induce renal tubular adenomas and adenocarcinomas in male rats regardless of the carrier vehicle. Chloroform has been studied in a wide variety of genotoxicity assays and has been found to give both positive and negative results.

The guideline value is based on extrapolation of the observed increase in kidney tumours in male rats exposed to chloroform in drinking-water for 2 years, although it is recognized that chloroform may induce tumours through a non-genotoxic mechanism. Using the linearized multistage model, a guideline value of 200 µg/litre vas calculated to correspond to an excess lifetime cancer risk of $10^{-5}$. This guideline value is supported by a 7.5-year study in dogs, in which a LOAEL of 15 mg/kg of body weight per day was observed for liver effects (applying an uncertainty factor of 1000 (100 for intra- and interspecies variation and 10 for the use of a LOAEL) and allocating 50% of the TDI to drinking-water).

Chlorinated acetic acids
The chlorinated acetic acids are oxidation by-products formed by the reaction of chlorine with organic material, such as humic or fulvic acids, present in water.

*Monochloroacetic acid*
Concentrations of monochloroacetic acid in chlorine-disinfected water are generally 1 µg/litre or less. In a recent 2-year bioassay in rats and mice, there was no

evidence of carcinogenicity. Available toxicity data are considered insufficient for deriving a guideline value.

*Dichloroacetic acid*

Dichloroacetic acid has been used pharmaceutically, as well as being a disinfection by-product. Concentrations in drinking-water in the United States of America of up to 80 µg/litre have been reported.

Dichloroacetic acid is readily absorbed following ingestion, rapidly metabolized to glyoxalate and oxalate, and excreted. In short- and long-term studies in laboratory animals, it induced neuropathy, decreases in body weight, testicular damage, and histopathological effects in the brain. Neuropathy was observed in one patient receiving therapeutic doses of dichloroacetate as a hypolipidaemic agent.

In several bioassays, dichloroacetate has been shown to induce hepatic tumours in mice. No adequate data on genotoxicity are available.

Because the evidence for the carcinogenicity of dichloroacetate is insufficient, a TDI of 7.6 µg/kg of body weight was calculated based on a NOAEL of 7.6 mg/kg of body weight per day for absence of effects on the liver in a 75-week study in mice and incorporating an uncertainty factor of 1000 (100 for intra- and interspecies variation and 10 for possible carcinogenicity). With an allocation of 20% of the TDI to drinking-water, the provisional guideline value is 50 µg/litre (rounded figure).

The guideline value is designated as provisional because the data are insufficient to ensure that the value is technically achievable. Difficulties in meeting a guideline value must never be a reason to compromise adequate disinfection.

*Trichloroacetic acid*

Trichloroacetic acid is used as a herbicide, as well as being a disinfection by-product. Concentrations in drinking-water of up to 100 µg/litre have been reported in the United States of America.

In short- and long-term studies in animal species, trichloroacetate has been shown to induce peroxisomal proliferation and increases in liver weight.

Trichloroacetate has been shown to induce tumours in the liver of mice. It has not been found to be mutagenic in *in vitro* assays. It has been reported to cause chromosomal aberrations.

Because the evidence for the carcinogenicity of trichloroacetic acid is restricted to one species, a TDI of 17.8 µg/kg of body weight was calculated based on a LOAEL of 178 mg/kg of body weight per day for an increase in liver weight in a 52-week study in mice and incorporating an uncertainty factor of 10 000 (100 for intra- and interspecies variation and 100 for the use of a slightly less-than-lifetime study, use of a LOAEL rather than a NOAEL, and possible carcinogenicity). A NOAEL in a 14-day study for the same effect was one-third of the LOAEL in the 52-week study. Based on a 20% allocation of the TDI to drinking-water, the provisional guideline value is 100 µg/litre (rounded figure).

The guideline value is designated as provisional because of the limitations of the available toxicological database and because there are inadequate data to judge whether the guideline value is technically achievable. Difficulties in meeting the guideline value must never be a reason for compromising adequate disinfection.

### Chloral hydrate (trichloroacetaldehyde)

Chloral hydrate is formed as a by-product of chlorination when chlorine reacts with humic acids. It has been found in drinking-water at concentrations of up to 100 µg/litre. It has been widely used as a sedative or hypnotic drug in humans at oral doses of up to 14 mg/kg of body weight.

The information available on the toxicity of chloral hydrate is limited, but effects on the liver have been observed in 90-day studies in mice. Chloral hydrate has been shown to be mutagenic in short-term tests *in vitro*, but it does not bind to DNA. It has been found to disrupt chromosome segregation in cell division.

A guideline value was calculated by applying an uncertainty factor of 10 000 (100 for intra- and interspecies variation, 10 for the short duration of the study, and 10 for the use of a LOAEL instead of a NOAEL) to the LOAEL of 16 mg/kg of body weight per day for liver enlargement from a 90-day drinking-water study in mice, to give a TDI of 1.6 µg/kg of body weight. With an allocation of 20% of the TDI to drinking-water, the provisional guideline value is 10 µg/litre (rounded figure). The guideline value is designated as provisional because of the limitations of the available database.

### Chloroacetones

1,1-Dichloroacetone is formed from the reaction between chlorine and organic precursors and has been detected in chlorinated drinking-water.

The toxicological data on 1,1-dichloroacetone are very limited, although studies with single doses indicate that it affects the liver.

There are insufficient data at present to permit the proposal of guideline values for 1,1-dichloroacetone or any of the other chloroacetones.

### Halogenated acetonitriles

Halogenated acetonitriles are formed from organic precursors during chlorination of drinking-water. Concentrations of dihalogenated acetonitriles in drinking-water range up to 40 µg/litre; reported levels of trichloroacetonitrile are less than 1 µg/litre. Halogenated acetonitriles may also be formed *in vivo* following ingestion of chlorinated water.

Halogenated acetonitriles are readily absorbed from the gastrointestinal tract and rapidly metabolized to single-carbon compounds, including cyanide. In 90-day studies, dibromoacetonitrile and dichloroacetonitrile induced decreases in body weight; specific target organs were not identified. Dichloroacetonitrile and trichloroacetonitrile have also been shown to be teratogenic in rats. No data on the effects of bromochloroacetonitrile in short- or long-term studies were available.

The carcinogenic potential of halogenated acetonitriles has not been investigated in long-term bioassays. IARC has concluded that all four halogenated acetonitriles are not classifiable as to their carcinogenicity to humans (Group 3).

Dichloroacetonitrile and bromochloroacetonitrile have been shown to be mutagenic in bacterial assays, whereas results for dibromoacetonitrile and trichloroacetonitrile were negative. All four of these halogenated acetonitriles induced sister chromatid exchange and DNA strand breaks and adducts in mammalian cells *in vitro* but were negative in the mouse micronucleus test.

*Dichloroacetonitrile*
For dichloroacetonitrile, a TDI of 15 μg/kg of body weight was calculated from a NOAEL of 15 mg/kg of body weight per day for fetal resorptions, decreases in fetal weight and size, and malformations of the cardiovascular, digestive, and urogenital systems in offspring in a teratology study in rats, incorporating an uncertainty factor of 1000 (100 for intra- and interspecies variation and 10 for the severity of the effects at doses above the NOAEL). This NOAEL is consistent with that observed for effects on body weight in a 90-day study in rats. Allocating 20% of the TDI to drinking-water, the provisional guideline value is 90 μg/litre. The guideline value is designated as provisional because of the limitations of the database (i.e., lack of long-term toxicity and carcinogenicity bioassays).

*Dibromoacetonitrile*
For dibromoacetonitrile, a TDI of 23 μg/kg of body weight was calculated from a NOAEL of 23 mg/kg of body weight per day for effects on body weight in a 90-day study in rats, incorporating an uncertainty factor of 1000 (100 for intra- and interspecies variation and 10 for the short duration of the study). Allocating 20% of the TDI to drinking-water, a provisional guideline value of 100 μg/litre (rounded figure) is calculated. The guideline value is designated as provisional because of the limitations of the database (i.e., lack of long-term toxicity and carcinogenicity bioassays).

*Bromochloroacetonitrile*
Available data are insufficient to serve as a basis for derivation of a guideline value for bromochloroacetonitrile.

*Trichloroacetonitrile*
For trichloroacetonitrile, a TDI of 0.2 µg/kg of body weight was calculated from a NOAEL of 1 mg/kg of body weight for decreases in fetal weight and viability and for cardiovascular and urogenital malformations in a teratology study in rats, incorporating an uncertainty factor of 5000 (100 for intra- and interspecies variation, 10 for severity of the effects at doses above the NOAEL, and 5 for limitations of the database, i.e., no 90-day study). Assuming a 20% allocation of the TDI to drinking-water, a provisional guideline value of 1 µg/litre (rounded figure) is derived. The guideline value is designated as provisional because of the limitations of the database (i.e., lack of long-term studies).

Cyanogen chloride
Cyanogen chloride is a by-product of chloramination. It is a reaction product of organic precursors with hypochlorous acid in the presence of ammonium ion. Concentrations detected in drinking-water treated with chlorine and chloramine were 0.4 and 1.6 µg/litre, respectively.

Cyanogen chloride is rapidly metabolized to cyanide in the body. There are few data on the oral toxicity of cyanogen chloride, and the guideline value is based, therefore, on cyanide.

A guideline value of 70 µg/litre for cyanide as total cyanogenic compounds is proposed (see pages 46–47).

Chloropicrin
Chloropicrin, or trichloronitromethane, is formed by the reaction of chlorine with humic and amino acids and with nitrophenols. Its formation is increased in the presence of nitrates. Limited data from the United States of America indicate that concentrations in drinking-water are usually less than 5 µg/litre.

Decreased survival and body weights have been reported following long-term oral exposure in laboratory animals. Chloropicrin has been shown to be mutagenic in bacterial tests and in *in vitro* assays in lymphocytes.

Because of the high mortality in a carcinogenesis bioassay and the limited number of end-points examined in the 78-week toxicity study, the available data were considered inadequate to permit the establishment of a guideline value for chloropicrin.

## 3.7 Monitoring

Practical implementation of water quality standards or guidelines requires the collection and analysis of samples. Both these operations present problems that, if not dealt with, may invalidate the conclusions of monitoring and undermine the usefulness of the guidelines. This section describes the main difficulties involved and outlines the approaches needed to deal with them. If sampling and analysis programmes are to provide valid information on water quality, it is vital that their objectives are defined clearly and unambiguously. In turn, therefore, it is essential that water quality guidelines should be defined as precisely as possible. The definition of the substances of interest and the numerical formulation of the guideline values are particularly important.

Many substances can exist in water in a variety of physicochemical forms or "species", the properties of which may differ markedly from each other. Analytical methods must be carefully selected so that all species of interest are determined, while forms of no concern are excluded. Therefore, all the substances specified in the water quality guidelines must be defined unambiguously; for this purpose, it should be assumed that the values recommended in these guidelines are for total concentrations, i.e., all forms of the substances present.

### 3.7.1 Design of a sampling programme

In order to assess the quality of potable water supplied to consumers, information is normally required over a given period (during which the quality may vary). The sampling programme should be designed to cover both random and systematic variations in water quality and to ensure that the collected samples are representative of the water quality throughout the whole distribution system. The frequency of sampling must be high enough to enable the programme to provide meaningful information while at the same time conserving sampling and analytical effort. However, the frequency of sampling may be reduced when there is evidence that particular substances are never present or where water supplies are obtained from sources with limited exposure to industrial, domestic, and agricultural wastes.

The type and magnitude of spatial and temporal variations in the concentration of water constituents will depend upon both their sources and their behaviour in the distribution and service systems.

Substances can be classified into two main types:

*Type 1.* Substances whose concentration is unlikely to vary during distribution. The concentration of these substances in the distribution system is largely governed by the concentration in the water going into the supply, and the substances do not undergo any reaction in the distribution system. Examples of such substances are arsenic, chloride, fluoride, hardness, pesticides, sodium, and total dissolved solids.

*Type 2.* Substances whose concentration may vary during distribution. These include:

— Substances whose concentration during distribution is dependent mainly on the concentration in the water going into the supply, but which

may participate in reactions (which change the concentration) within the distribution system. Examples are aluminium, chloroform, iron, manganese and hydrogen ion (pH).
— Substances for which the distribution system provides the main source, such as benzo[*a*]pyrene, copper, lead, and zinc.

This classification applies only to piped water supplies. In all other types of supply, water constituents should be regarded as type 1 substances.

The same substance may belong to different classes in different distribution systems.

## *Frequency of appraisal*

Frequent sampling and appraisal are necessary for microbiological constituents, but sampling and analysis for the control of health-related organic and inorganic compounds in drinking-water are required less often. A thorough appraisal should be made when any new water source comes into service and immediately following any major change in the treatment processes. Subsequently, samples should be analysed periodically, the frequency being determined by local circumstances. In addition, local information on changes in the catchment area (especially agricultural and industrial activities) is important and can be used to predict possible contamination problems and, consequently, the need for more frequent monitoring of specific compounds.

The subject of frequency of appraisal of drinking-water for evaluation of aesthetic qualities cannot be generalized. Some constituents, for example sodium or chloride, are in the drinking-water at the source, and others are added during the water treatment processes. Other characteristics and constituents, such as taste, iron, zinc, etc., may vary considerably as a result of other considerations or in relation to the type of distribution system and the prevalence of corrosion problems. Obviously, for some constituents and characteristics the appraisal will need to be fairly frequent, whereas for others, where the levels show little variation, less frequent determination will be sufficient.

## *Sampling locations*

The exact sites for sampling need to be chosen carefully to provide samples that are representative of the whole system or of the particular problem area. Exact recommendations cannot be given on the selection of the correct site because of the complexities involved; sample locations are best chosen using local knowledge concerning the specific problems, the water source, and the distribution system.

For type 1 substances, it is generally sufficient to sample only the water going into the supply. Where two or more water sources with different concentrations of a type 1 substance are feeding the same distribution network, some additional sampling may be required within the distribution system.

The concentrations of type 2 substances are liable to change between the supply points and consumers' taps. Many interconnected processes may occur (e.g., corrosion of pipes, deposition of solids, reactions between substances in the water), which necessitate the collection of samples from consumers' taps. The selection of taps cannot be made on a general basis and must rely on consideration of the particular circumstance involved. However, two extreme sampling strategies may be distinguished: (i) taps selected on a wholly random basis; (ii) taps selected systematically on the basis of knowledge of factors affecting the substance of interest.

The nature and magnitude of spatial variations in quality and the monitoring objectives will determine which of these approaches (or a combination) is most appropriate. Random sampling is usually desirable when the spatial variations in quality are completely random, but it may not be ideal if there are systematic differences in quality between different parts of the distribution system. For lead, for example, random sampling might not be appropriate in a distribution system in which only 1% of the service and domestic plumbing pipes are made of lead. On the other hand, complete reliance on systematic sampling may be inappropriate. If random sampling is decided upon, it is important that the sample points should be selected on a truly random basis, care being taken that certain locations are not sampled regularly because of convenience or ease of access.

### *Sampling times*

Raw water quality, the efficiency of treatment processes, and the effects of the distribution system on drinking-water quality will all vary with time.

For type 1 substances, analysis of the water going into the supply usually provides an appropriate basis for monitoring. The principal factors that determine the times and frequency of sampling are therefore the concentration of the substance of interest, its variation, and the extent, if any, to which it is affected by treatment.

The concentrations of type 2 substances are affected by many processes and therefore tend to show complex and erratic variations with time. Each situation (substance, distribution system, information need) will require individual examination. The objectives of monitoring will greatly affect the choice of sampling times.

If temporal variations are completely random, the time of sampling is unimportant. Statistical estimation of the number of samples to be taken from a particular tap over a given period can, in principle, be made in such situations, but problems arise if systematic variations occur.

## 3. CHEMICAL ASPECTS

When there are rapid changes in water quality, the actual time span over which the sample is collected can significantly affect the analytical results. A composite sample, collected over a period of time, will give a time-weighted average value, whereas a single sample will give values highly dependent on cyclic and random variations. Continuous monitoring devices may be useful, but these are not generally available for all the variables of interest.

Sampling locations and times should be chosen jointly, as there is a limit to the amount of sampling and analysis that can be carried out. Two extreme strategies are: (1) to sample many taps, each on only one or a few occasions, and (2) to sample fewer taps, but each more frequently. It should be noted that too frequent sampling will produce unnecessary data and will considerably increase the cost.

The relative magnitudes of spatial and temporal variations will clearly be an important factor in selecting the strategy. Where spatial variations predominate, a greater effort will generally be directed to strategy (1) than to strategy (2), and vice versa.

### Monitoring to ensure compliance

If limits established in national legislation for type 2 substances are regarded as concentrations that must not be exceeded at any time or place, designing a sampling programme becomes extremely difficult. In the case of type 1 substances, for which monitoring at perhaps only one or a few locations is necessary, the difficulties are fewer, but some problems do still arise.

If continuous monitoring is not possible, a number of individual samples should be taken for analysis and the quality of the supply at other times inferred statistically from the results. It is difficult, however, to estimate maximum values from such data (in particular because the nature of the statistical distribution of sample concentrations will often not be known), and the estimated maxima will be subject to relatively large uncertainties. In these circumstances, alternative criteria for judging compliance will be needed. For example, the criterion of compliance could be defined as follows: "That $x$% of all possible samples (i.e., $x$% of the statistical population) do not exceed the limit." However, because only a limited number of results will be available, uncertainties in estimating such a percentage must be recognized. The risks of drawing false conclusions must be reduced to acceptable levels by the choice of an appropriate number of samples and of appropriate analytical error limits. Of course, other criteria — for example, based on the mean concentration of the substance — could be employed.

In addition to the statistical approach to judging compliance, attention must also be paid to the choice of sampling times (and locations, in the case of type 2 substances) in relation to the behaviour of the particular substance in the distribution system. For example, in the case of lead, a variety of sample types is possible, such as first-draw samples (i.e., samples taken after overnight stagnation), random daytime samples, flushed samples, etc. First-draw samples will have the highest lead concentrations but are the least convenient to collect. Flushed samples, on the other hand, give the most consistent values but reflect the minimum exposure of the water to lead. The random daytime samples, although most truly reflecting the water that the consumer drinks, give the most variable levels, and so it is necessary to collect more samples to determine the mean level of exposure. Considerations similar to those outlined above will apply to other type 2 substances, although the spatial and temporal variations are likely, of course, to follow different patterns.

Finally, when considering criteria for judging compliance with a limit, attention must be given to the area and time over which the assessment of compliance will be made. Generally, the area should be based on the individual water supply system, although subdivision of water supply systems may be useful if the distribution materials differ markedly in different parts of the system. In some circumstances, it may be desirable to increase the number of samples taken in proportion to the size of the population served to avoid the risks of drawing false conclusions concerning compliance.

### 3.7.2 Sample collection

Samples should fulfil two conditions: (1) the water entering the sample container should be a representative sample, and (2) the concentration of the substance being determined should not change between sampling and analysis.

#### *Consumers' taps*

When all or part of the water emerging from a tap is collected, the concentration of a substance of interest may be affected by two main factors: the flow rate from the tap and the volume collected. Substances of type 1 are not usually affected by these factors; however, for type 2 substances, two fundamental problems arise:

- If the flow-rate normally used by the consumer is also used for sampling, there may well be difficulties in comparing the qualities observed at different taps sampled at different flow rates. On the other hand, if a standardized flow rate is adopted to reduce this problem, the observed qualities may then not reflect the quality of water as used by the consumer.

- When the samples are taken at times of rapid or systematic change in water quality, the volume of the sample collected may affect the observed quality. In this case, a practical solution is to specify the particular sample volume to be collected.

## 3. CHEMICAL ASPECTS

### *Sample stability*

The concentrations of the substances to be determined in a sample may change between sampling and analysis as a result of (1) external contamination during the collection of the sample, (2) contamination from the container, or (3) chemical, physical, and biological processes in the sample.

Serious errors can occur unless appropriate precautions are taken, but, generally, standard or recommended methods of analysis are designed to avoid contamination from the sample container and to minimize concentration changes during storage. Moreover, the method of sample preservation will often be determined by the analytical method employed. Tests should nevertheless be carried out to check that the concentration of the substance being determined does not change unacceptably during the period between sample collection and analysis.

### 3.7.3 Analysis

When a representative sample of water is analysed for a substance of interest, the accuracy of the result depends entirely on what errors arise during analysis.

International laboratory studies have shown that in certain laboratories serious errors of analysis occur, sometimes as large as several hundred percent. Commonly, this analytical error is greatest for substances that are present at low concentrations. Quality control should be a fundamental part of any programme of sampling and analysis, especially when the results of the work are to be compared with numerical standards or guidelines. Suitable analytical procedures are generally available to reach the required standards of accuracy; the practical problem is to ensure their correct application. In some countries, there will also be problems related to the availability of the necessary equipment. If these problems are to be avoided, it is important that the maximum total tolerable error for each substance should be decided upon on the basis of the information required from the monitoring (or identification) work, and that appropriate analytical methods are employed and properly applied so that the required accuracy is achieved.

Various general aspects related to these two points are considered in the following sections.

### *Defining the required accuracy*

The accuracy required of an analytical procedure is, in principle, governed by the objectives of the programme of sampling and analysis, which will be different in different circumstances. Consequently, a generally applicable definition of the required accuracy cannot be given, and attention is restricted here to consideration of four points of particular importance.

- The accuracy required should be defined in an explicit, quantitative manner, so that unambiguous criteria are available for the selection of suitable analytical methods. In the absence of such criteria, a laboratory's approach to the selection of methods may be governed by other factors (e.g., speed, cost), to the detriment of accuracy.

- As the target for the accuracy of any analysis is made more stringent, the time and effort required (and therefore the cost) will increase — often disproportionately to the improvement in accuracy. A frequent and costly practice is to set the limit of accuracy on the basis of analytical and statistical considerations only without considering the real meaning of a given error. For some substances at low concentrations, even an error of $\pm 50\%$ may have no sanitary or health significance. The setting of needlessly stringent targets should therefore be avoided.

- Many of the substances considered in these guidelines may be present at very low concentrations, and therefore the limit of detection is often likely to be the single most important criterion in selecting a method of analysis. It is essential that the smallest concentration of interest should be identified. This concentration will, in general, be considered as the required limit of detection. It may be useful, therefore, to set the required limit of detection to 20% of the recommended guideline value.

- Careful consideration should be given to the manner of expressing target accuracy. The target accuracy should be expressed in terms of the maximum tolerable total error with a defined confidence level.

### Selecting suitable analytical methods

Various collections of "standard" or "recommended" methods for water analysis are published by a number of national and international agencies. It is often thought that adequate analytical accuracy can be achieved without problems provided that all laboratories use the same standard method. Experience shows that this is not the case, as a variety of extraneous factors may affect the accuracy of the results. Examples include reagent purity, apparatus type and performance, degree of modification of the method in a particular laboratory, and the skill and care of the analyst. These factors are likely to vary, both between laboratories and over time in an individual laboratory. Moreover, the accuracy that can be achieved with a particular method frequently depends upon the nature and composition of the sample. It is not essential to use standard methods except in the case of "non-specific" variables such as taste and odour, colour, and turbidity. In these cases, the result is determined by the method employed, and it is necessary for all laboratories to use identical methods if comparable results are to be obtained.

A number of considerations are important in selecting analytical methods:

- The overriding consideration is that the method chosen can result in the required accuracy. Other factors, such as speed and convenience, should be considered only in selecting among methods that meet this primary criterion.

- There are a number of markedly different procedures for measuring and reporting the errors to which methods are subject. This needlessly complicates and prejudices the effectiveness of method selection, and suggestions for standardizing such procedures have been made. It is desirable that details of all analytical methods are published together with performance characteristics that can be interpreted unambiguously.

- If the analytical results from one laboratory are to be compared with those from others and/or with a numerical standard, it is obviously preferable for them not to have any associated systematic error. In practice, this is not possible, but each laboratory should select methods whose systematic errors have been thoroughly evaluated and shown to be acceptably small.

### *Analytical quality control*

Whichever method is chosen, appropriate analytical quality control procedures must be implemented to ensure that the results produced are of adequate accuracy. Because of the wide range of substances, methods, equipment, and accuracy requirements likely to be involved in the monitoring of drinking-water, many detailed, practical aspects of analytical quality control are concerned. These are beyond the scope of this publication, which can give only an idea of the approach involved.

Before analysing samples by the chosen method, preliminary tests should be conducted by each laboratory to provide estimates of its precision (random error of the results). The routine analysis of samples (accompanied by regular checks of precision) can begin when the results from the preliminary tests have acceptably small errors. These preliminary tests can, and should, check certain sources of systematic error, but this is usually very difficult for a routine laboratory. This emphasizes the need for sound selection of methods initially, and also for another form of analytical quality control, namely, interlaboratory testing. Such testing is usually the best single approach to checking systematic error but should be undertaken only after satisfactory completion of preliminary tests of precision. There may be some difficulty in implementing an analytical quality control programme if the coordinating laboratory has to deal with a large number of other laboratories or if the laboratories are far apart. A hierarchical structure of coordinating and participating laboratories allows any such difficulty to be overcome.

# 4.
# Radiological aspects

## 4.1 Introduction

The guideline levels for radioactivity in drinking-water recommended in the first edition of *Guidelines for drinking-water quality* in 1984 were based on the data available at that time on the risks of exposure to radiation sources. Since then, additional information has become available on the health consequences of exposure to radiation, risk estimates have been reviewed, and the recommendations of the International Commission on Radiological Protection (ICRP) have been revised. This new information has been taken into account in the preparation of the recommendations in this chapter.

The purpose of these recommendations for radioactive substances in drinking-water is to guide the competent authorities in determining whether the water is of an appropriate quality for human consumption.

### 4.1.1 Environmental radiation exposure

Environmental radiation originates from a number of naturally occurring and man-made sources. The United Nations Scientific Committee on the Effects of Atomic Radiation (UNSCEAR) has estimated that exposure to natural sources contributes more than 98% of the radiation dose to the population (excluding medical exposure). There is only a very small contribution from nuclear power production and nuclear weapons testing. The global average human exposure from natural sources is 2.4 mSv/year. There are large local variations in this exposure depending on a number of factors, such as height above sea level, the amount and type of radionuclides in the soil, and the amount taken into the body in air, food, and water. The contribution of drinking-water to the total exposure is very small and is due largely to naturally occurring radionuclides in the uranium and thorium decay series.

Levels of natural radionuclides in drinking-water may be increased by a number of human activities. Radionuclides from the nuclear fuel cycle and from medical and other uses of radioactive materials may enter drinking-water supplies; the contributions from these sources are normally limited by regulatory control of the source or practice, and it is through this regulatory mechanism that remedial

# 4. RADIOLOGICAL ASPECTS

action should be taken in the event that such sources cause concern by contaminating drinking-water.

## 4.1.2 Potential health consequences of radiation exposure

Exposure to ionizing radiation, whether natural or man-made, can cause two kinds of health effects. Effects for which the severity of the damage caused is proportional to the dose, and for which a threshold exists below which the effect does not occur, are called "deterministic" effects. Under normal conditions, the dose received from natural radioactivity and routine exposures from regulated practices is well below the threshold levels, and therefore deterministic effects are not relevant to these recommendations.

Effects for which the probability of occurrence is proportional to dose are known as "stochastic" effects, and it is assumed that there is no threshold below which they do not occur. The main stochastic effect of concern is cancer.

Because different types of radiation have different biological effectiveness and different organs and tissues in the body have different sensitivities to radiation, the ICRP has introduced radiation and tissue-weighting factors to provide a measure of equal effect. The sum of the doubly weighted dose received by all the tissues and organs of the body gives a measure of the total harm and is referred to as the effective dose. Moreover, radionuclides taken into the body may persist, and, in some cases, the resulting exposure may extend over many months or years. The committed effective dose is a measure of the total effective dose incurred over a lifetime following the intake of a radionuclide. It is this measure of exposure that is relevant to the present discussion; in what follows, the term "dose" refers to the committed effective dose, which is expressed in sieverts (Sv). The risk of adverse health consequences from radiation exposure is a function of the total dose received from all sources. A revised estimate of the risk (i.e., the mathematical expectation) of a lifetime fatal cancer for the general population has been estimated by the ICRP to be $5 \times 10^{-2}$ per sievert. (This does not include a small additional health risk from non-fatal cancers or hereditary effects.)

## 4.1.3 Recommendations

- The recommended reference level of committed effective dose is 0.1 mSv from 1 year's consumption of drinking-water. This reference level of dose represents less than 5% of the average effective dose attributable annually to natural background radiation.
- Below this reference level of dose, the drinking-water is acceptable for human consumption, and any action to reduce the radioactivity is not necessary.

- For practical purposes, the recommended guideline activity concentrations are 0.1 Bq/litre for gross alpha and 1 Bq/litre for gross beta activity.

The recommendations apply to routine operational conditions of existing or new water supplies. They do not apply to a water supply contaminated during an emergency involving the release of radionuclides into the environment. Guidelines covering emergencies are available elsewhere (see Bibliography).

The recommendations do not differentiate between natural and man-made radionuclides.

## 4.2 Application of the reference level of dose

For practical purposes, the reference level of dose needs to be expressed as an activity concentration of radionuclides in drinking-water.

The dose to a human from radioactivity in drinking-water is dependent not only on intake but also on metabolic and dosimetric considerations. The guideline activity concentrations assume an intake of total radioactive material from the consumption of 2 litres of water per day for 1 year and are calculated on the basis of the metabolism of an adult. The influence of age on metabolism and variations in consumption of drinking-water do not require modification of the guideline activity concentrations, which are based on a lifetime exposure and provide an appropriate margin of safety. Metabolic and dosimetric considerations have been included in the development of dose conversion factors, expressed as sieverts per becquerel, which relate a dose expressed in sieverts to the quantity (in becquerels) of radioactive material ingested.

Examples of radionuclide concentrations (reference concentrations) corresponding to the reference level of dose, 0.1 mSv/year, are given in Table 8. These concentrations have been calculated using the dose conversion factors of the United Kingdom National Radiological Protection Board from the formula:

reference concentration (Bq/litre)

$$= \frac{1 \times 10^{-4} \text{ (Sv/year)}}{730 \text{ (litre/year)} \times \text{dose conversion factor (Sv/Bq)}}$$

$$= \frac{1.4 \times 10^{-7} \text{ (Sv/litre)}}{\text{dose conversion factor (Sv/Bq)}}$$

The previous guidelines recommended the use of an average gross alpha and gross beta activity concentration for routine screening. These were set at 0.1 Bq/litre and 1 Bq/litre, respectively. The doses associated with these levels of gross alpha and gross beta activity for selected radionuclides are shown in Table 9. For some radionuclides, such as $^{226}$Ra and $^{90}$Sr, the associated dose is much lower than

## 4. RADIOLOGICAL ASPECTS

*Table 8. Activity concentration of various radionuclides in drinking-water corresponding to a dose of 0.1 mSv from 1 year's intake*

| Radionuclide[a] | Dose conversion factor (Sv/Bq)[b] | Calculated rounded value (Bq/litre) |
|---|---|---|
| $^{3}$H | $1.8 \times 10^{-11}$ | 7800 |
| $^{14}$C | $5.6 \times 10^{-10}$ | 250 |
| $^{60}$Co | $7.2 \times 10^{-9}$ | 20 |
| $^{89}$Sr | $3.8 \times 10^{-9}$ | 37 |
| $^{90}$Sr | $2.8 \times 10^{-8}$ | 5 |
| $^{129}$I | $1.1 \times 10^{-7}$ | 1 |
| $^{131}$I | $2.2 \times 10^{-8}$ | 6 |
| $^{134}$Cs | $1.9 \times 10^{-8}$ | 7 |
| $^{137}$Cs | $1.3 \times 10^{-8}$ | 10 |
| $^{210}$Pb | $1.3 \times 10^{-6}$ | 0.1 |
| $^{210}$Po | $6.2 \times 10^{-7}$ | 0.2 |
| $^{224}$Ra | $8.0 \times 10^{-8}$ | 2 |
| $^{226}$Ra | $2.2 \times 10^{-7}$ | 1 |
| $^{228}$Ra | $2.7 \times 10^{-7}$ | 1 |
| $^{232}$Th | $1.8 \times 10^{-6}$ | 0.1 |
| $^{234}$U | $3.9 \times 10^{-8}$ | 4 |
| $^{238}$U | $3.6 \times 10^{-8}$ | 4 |
| $^{239}$Pu | $5.6 \times 10^{-7}$ | 0.3 |

[a] For $^{40}$K, see page 118. For $^{222}$Rn, see section 4.2.3.
[b] Values from National Radiological Protection Board, *Committed equivalent organ doses and committed effective doses from intakes of radionuclides*. Chilton, Didcot, 1991.

0.1 mSv per year. It can also be seen from this table that, if certain radionuclides, such as $^{232}$Th, $^{228}$Ra, or $^{210}$Pb, are singly responsible for 0.1 Bq/litre for gross alpha activity or 1 Bq/litre for gross beta activity, then the reference level of dose of 0.1 mSv per year would be exceeded. However, these radionuclides usually represent only a small fraction of the gross activity. In addition, an elevated activity concentration of these radionuclides would normally be associated with high activities from other radionuclides. This would elevate the gross alpha or gross beta activity concentration above the investigation level and provoke specific radionuclide analysis. Therefore, the values of 0.1 Bq/litre for gross alpha activity and 1 Bq/litre for gross beta activity continue to be recommended as screening levels for drinking-water, below which no further action is required.

Radionuclides emitting low-energy beta particles, such as $^{3}$H and $^{14}$C, and some gaseous or volatile radionuclides, such as $^{222}$Rn and $^{131}$I, will not be detected by standard methods of measurement. The values for average gross alpha and beta activities do not include such radionuclides, so that if their presence is suspected, special sampling techniques and measurements should be used.

**Table 9. Examples of the doses arising from 1 year's consumption of drinking-water containing any of the given alpha-emitting radionuclides at an activity concentration of 0.1 Bq/litre or of the given beta-emitting radionuclides at an activity concentration of 1 Bq/litre**[a]

| Radionuclide | Dose (mSv) |
|---|---|
| Alpha emitters (0.1 Bq/litre) | |
| $^{210}$Po | 0.045 |
| $^{224}$Ra | 0.006 |
| $^{226}$Ra | 0.016 |
| $^{232}$Th | 0.130 |
| $^{234}$U | 0.003 |
| $^{238}$U | 0.003 |
| $^{239}$Pu | 0.04 |
| Beta emitters (1 Bq/litre) | |
| $^{60}$Co | 0.005 |
| $^{89}$Sr | 0.003 |
| $^{90}$Sr | 0.020 |
| $^{129}$I | 0.080 |
| $^{131}$I | 0.016 |
| $^{134}$Cs | 0.014 |
| $^{137}$Cs | 0.009 |
| $^{210}$Pb | 0.95 |
| $^{228}$Ra | 0.20 |

[a] Appropriate dose conversion factors taken from National Radiological Protection Board, *Committed equivalent organ doses and committed effective doses from intakes of radionuclides*, Chilton, Didcot, 1991.

It should not necessarily be assumed that the reference level of dose has been exceeded simply because the gross beta activity concentration approaches or exceeds 1 Bq/litre. This situation may well result from the presence of the naturally occurring radionuclide $^{40}$K, which makes up about 0.01 % of natural potassium. The absorption of the essential element potassium is under homoeostatic control and takes place mainly from ingested food. Thus, the contribution to dose from the ingestion of $^{40}$K in drinking-water, with its relatively low dose conversion factor ($5 \times 10^{-9}$ Sv/Bq), will be much less than that of many other beta-emitting radionuclides. This situation will be clarified by the identification of the specific radionuclides in the sample.

### 4.2.1 Analytical methods

The International Organization for Standardization (ISO) has published standard methods for determining gross alpha and gross beta activity concentrations

# 4. RADIOLOGICAL ASPECTS

in water. Although the detection limits depend on the radionuclides present, the dissolved solids in the sample, and the counting conditions, the recommended levels for gross alpha and gross beta activity concentrations should be above the limits of detection. The ISO detection limit for gross alpha activity based on $^{239}$Pu is 0.04 Bq/litre, while that for gross beta activity based on $^{137}$Cs is between 0.04 and 0.1 Bq/litre.

For analyses of specific radionuclides in drinking-water, there are general compendium sources in addition to specific methods in the technical literature (see Bibliography).

### 4.2.2 Strategy for assessing drinking-water

If either the gross alpha activity concentration of 0.1 Bq/litre or the gross beta activity concentration of 1 Bq/litre is exceeded, then the specific radionuclides should be identified and their individual activity concentrations measured. From these data, a dose estimate for each radionuclide should be made and the sum of these doses determined. Where the following additive formula is satisfied, no further action is required:

$$\sum_i \frac{C_i}{RC_i} \leq 1$$

where $C_i$ is the measured activity concentration of radionuclide $i$ and $RC_i$ is the reference activity concentration of radionuclide $i$ that, at an intake of 2 litres per day for 1 year, will result in a committed effective dose of 0.1 mSv (see Table 8).

If alpha-emitting radionuclides with high dose conversion factors are suspected, this additive formula may also be invoked when the gross alpha and gross beta activity screening values of 0.1 Bq/litre and 1 Bq/litre are approached. Where the sum exceeds unity for a single sample, the reference level of dose of 0.1 mSv would be exceeded only if the exposure to the same measured concentrations were to continue for a full year. Hence, such a sample does not in itself imply that the water is unsuitable for consumption and should be regarded only as a level at which further investigation, including additional sampling, is needed.

The options available to the competent authority to reduce the dose should then be examined. Where remedial measures are contemplated, any strategy considered should first be justified (in the sense that it achieves a positive net benefit) and then optimized in accordance with the recommendations of ICRP in order to produce the maximum net benefit. The application of these recommendations is summarized in Fig. 1.

**Fig. 1. Application of recommendations on radionuclides in drinking-water based on an annual reference level of dose of 0.1 mSv**

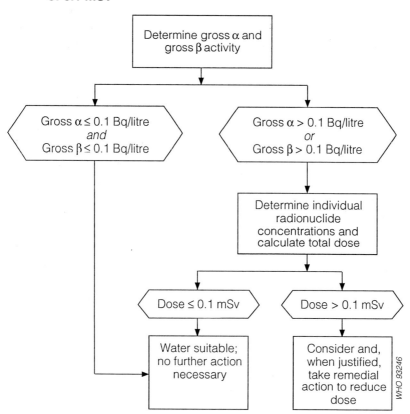

### 4.2.3 Radon

There are difficulties in applying the reference level of dose to derive activity concentrations of $^{222}$Rn in drinking-water. These difficulties arise from the ease with which radon is released from water during handling and the importance of the inhalation pathway. Stirring and transferring water from one container to another will liberate dissolved radon. Water that has been left to stand will have reduced radon activity, and boiling will remove radon completely. As a result, it is important that the form of water consumed is taken into account in assessing the dose from ingestion. Moreover, the use of water supplies for other domestic uses will increase the levels of radon in the air, thus increasing the dose from inhalation. This dose depends markedly on the form of domestic usage and housing construction. The form of water intake, the domestic use of water, and the construc-

## 4. RADIOLOGICAL ASPECTS

tion of houses vary widely throughout the world. It is therefore not possible to derive an activity concentration for radon in drinking-water that is universally applicable.

The global average dose from inhalation of radon from all sources is about 1 mSv/year, which is roughly half of the total natural radiation exposure. In comparison, the global dose from ingestion of radon in drinking-water is relatively low. In a local situation, however, the risk from inhalation and ingestion may be about equal. Because of this and because there may be other sources of radon gas entry to a house, ingestion cannot be considered in isolation from inhalation exposures.

All these factors should be assessed on a regional or national level by the appropriate authorities, in order to determine whether a reference level of dose of 0.1 mSv is appropriate for that region, and to determine an activity concentration that may be used to assess the suitability of the water supply. These judgements should be based not only on the ingestion and inhalation exposures resulting from the supply of water, but also on the inhalation doses from other radon sources in the home. In these circumstances, it would appear necessary to adopt an integrated approach and assess doses from all radon sources, especially to determine the optimum action to be undertaken where some sort of intervention is deemed necessary.

# 5.
# Acceptability aspects

## 5.1 Introduction

The most undesirable constituents of drinking-water are undoubtedly those that are capable of having a direct impact on public health and for which guideline values have been developed. The management of these substances is in the hands of organizations responsible for the provision of the supply, and it is up to these organizations to instil in their consumers the confidence that this task is being undertaken with responsibility and efficiency.

To a large extent, consumers have no means of judging the safety of their drinking-water themselves, but their attitude towards their water supply and their water suppliers will be affected to a considerable extent by the aspects of water quality that they are able to perceive with their own senses. It is natural, therefore, for consumers to regard with grave suspicion water that appears dirty or discoloured or that has an unpleasant taste or smell, even though these characteristics may not in themselves be of any direct consequence to health.

The provision of drinking-water that is not only safe but also pleasing in appearance, taste, and odour is a matter of high priority. The supply of water that is unsatisfactory in this respect will undermine the confidence of consumers, leading to complaints and possibly the use of water from less safe sources. It can also result in the use of bottled water, which is expensive, and home treatment devices, some of which can have adverse effects on water quality.

The acceptability of drinking-water to consumers can be influenced by many different constituents; most of the substances for which guideline values have been set, and which also affect the taste or odour of water, have been referred to already (see section 3.6). There are a number of other water constituents that are of no direct consequence to health at the concentrations at which they normally occur in water but which nevertheless may be objectionable to consumers for various reasons.

The concentration at which such constituents are offensive to consumers is dependent on individual and local factors, including the quality of the water to which the community is accustomed and a variety of social, economic, and cultural considerations. Under these circumstances, it is inappropriate to set guide-

# 5. ACCEPTABILITY ASPECTS

line values specific to substances that affect the acceptability of water to consumers but which are not directly relevant to health.

In the following summary statements, reference is made to levels likely to give rise to complaints from consumers. These are not precise numbers, and problems may occur at lower or much higher levels, depending on individual and local circumstances.

## 5.2 Summary statements

### 5.2.1 Physical parameters

#### Colour

The colour of drinking-water is usually due to the presence of coloured organic matter (primarily humic and fulvic acids) associated with the humus fraction of soil. Colour is strongly influenced by the presence of iron and other metals, either as natural impurities or as corrosion products. It may also result from the contamination of the water source with industrial effluents and may be the first indication of a hazardous situation. The source of colour in a water supply should be investigated, particularly if a substantial change takes place.

Colours above 15 TCU (true colour units) can be detected in a glass of water by most people. Colours below 15 TCU are usually acceptable to consumers, but acceptability may vary according to local circumstances.

No health-based guideline value is proposed for colour in drinking-water.

#### Taste and odour

Taste and odour originate from natural and biological sources or processes (e.g., aquatic microorganisms), from contamination by chemicals, or as a by-product of water treatment (e.g., chlorination). Taste and odour may also develop during storage and distribution.

Taste and odour in drinking-water may be indicative of some form of pollution or of malfunction during water treatment or distribution. The cause of tastes and odours should be investigated and the appropriate health authorities should be consulted, particularly if there is a sudden or substantial change. An unusual taste or odour might be an indication of the presence of potentially harmful substances.

The taste and odour of drinking-water should not be offensive to the consumer. However, there is an enormous variation in the level and quality of taste and odour that are regarded as acceptable.

No health-based guideline value is proposed for taste and odour.

#### Temperature

Cool water is generally more palatable than warm water. High water temperature enhances the growth of microorganisms and may increase taste, odour, colour, and corrosion problems.

## Turbidity

Turbidity in drinking-water is caused by particulate matter that may be present as a consequence of inadequate treatment or from resuspension of sediment in the distribution system. It may also be due to the presence of inorganic particulate matter in some ground waters.

High levels of turbidity can protect microorganisms from the effects of disinfection and can stimulate bacterial growth. In all cases where water is disinfected, therefore, the turbidity must be low so that disinfection can be effective. The impact of turbidity on disinfection efficiency is discussed in more detail in Chapter 6.

The appearance of water with a turbidity of less than 5 nephelometric turbidity units is usually acceptable to consumers, although this may vary with local circumstances. However, because of its microbiological effects, it is recommended that turbidity be kept as low as possible. No health-based guideline value for turbidity has been proposed.

### 5.2.2 Inorganic constituents

## Aluminium

The presence of aluminium at concentrations in excess of 0.2 mg/litre often leads to consumer complaints as a result of deposition of aluminium hydroxide floc in distribution systems and the exacerbation of discoloration of water by iron; concentrations between 0.1 and 0.2 mg/litre may give rise to these problems in some circumstances.

Available evidence does not support the derivation of a health-based guideline value for aluminium in drinking-water (see page 39).

## Ammonia

The threshold odour concentration of ammonia at alkaline pH is approximately 1.5 mg/litre, and a taste threshold of 35 mg/litre has been proposed for the ammonium cation.

Ammonia is not of immediate health relevance, and no health-based guideline value has been proposed (see page 40).

## Chloride

High concentrations of chloride give an undesirable taste to water and beverages. Taste thresholds for the chloride anion depend on the associated cation and are in the range of 200–300 mg/litre for sodium, potassium, and calcium chloride. Consumers can become accustomed to concentrations in excess of 250 mg/litre.

No health-based guideline value is proposed for chloride in drinking-water (see page 45).

### Copper

The presence of copper in a water supply may interfere with the intended domestic uses of the water. Copper in public water supplies increases the corrosion of galvanized iron and steel fittings. Staining of laundry and sanitary ware occurs at copper concentrations above 1 mg/litre. At levels above 5 mg/litre, it also imparts a colour and an undesirable bitter taste to water.

Although copper can give rise to taste problems, the taste should be acceptable at the health-based provisional guideline value (see page 46).

### Hardness

Public acceptability of the degree of hardness of water may vary considerably from one community to another, depending on local conditions. The taste threshold for the calcium ion is in the range 100–300 mg/litre, depending on the associated anion, and the taste threshold for magnesium is probably less than that for calcium. In some instances, a water hardness in excess of 500 mg/litre is tolerated by consumers.

Depending on the interaction of other factors, such as pH and alkalinity, water with a hardness above approximately 200 mg/litre may cause scale deposition in the distribution system and will result in excessive soap consumption and subsequent "scum" formation. On heating, hard waters form deposits of calcium carbonate scale. Soft water, with a hardness of less than 100 mg/litre, may, on the other hand, have a low buffer capacity and so be more corrosive for water pipes (see section 6.6).

No health-based guideline value has been proposed for hardness (see page 48).

### Hydrogen sulfide

The taste and odour thresholds of hydrogen sulfide in water are estimated to be between 0.05 and 0.1 mg/litre. The "rotten eggs" odour of hydrogen sulfide is particularly noticeable in some ground waters and in stagnant drinking-water in the distribution system, as a result of oxygen depletion and the subsequent reduction of sulfate by bacterial activity.

Sulfide is oxidized rapidly to sulfate in well-aerated water, and hydrogen sulfide levels in oxygenated water supplies are normally very low. The presence of hydrogen sulfide in drinking-water can be easily detected by the consumer and requires immediate corrective action. It is unlikely that a person could consume

### Iron

Anaerobic ground water may contain ferrous iron at concentrations of up to several milligrams per litre without discoloration or turbidity in the water when directly pumped from a well. On exposure to the atmosphere, however, the ferrous iron oxidizes to ferric iron, giving an objectionable reddish-brown colour to the water.

Iron also promotes the growth of "iron bacteria", which derive their energy from the oxidation of ferrous iron to ferric iron and in the process deposit a slimy coating on the piping.

At levels above 0.3 mg/litre, iron stains laundry and plumbing fixtures. There is usually no noticeable taste at iron concentrations below 0.3 mg/litre, although turbidity and colour may develop. Iron concentrations of 1–3 mg/litre can be acceptable for people drinking anaerobic well-water.

No health-based guideline value is proposed for iron (see page 48).

### Manganese

Although manganese concentrations below 0.1 mg/litre are usually acceptable to consumers, this may vary with local circumstances. At levels exceeding 0.1 mg/litre, manganese in water supplies stains sanitary ware and laundry and causes an undesirable taste in beverages. The presence of manganese in drinking-water, like that of iron, may lead to the accumulation of deposits in the distribution system. Even at a concentration of 0.02 mg/litre, manganese will often form a coating on pipes, which may slough off as a black precipitate. In addition, certain nuisance organisms concentrate manganese and give rise to taste, odour, and turbidity problems in distributed water.

Although concentrations below 0.1 mg/litre are usually acceptable to consumers, this may vary with local circumstances. The provisional health-based guideline value for manganese is 5 times higher than this acceptability threshold of 0.1 mg/litre (see page 50).

### Dissolved oxygen

The dissolved oxygen content of water is influenced by the raw water temperature, composition, treatment, and any chemical or biological processes taking place in the distribution system. Depletion of dissoved oxygen in water supplies can encourage the microbial reduction of nitrate to nitrite and sulfate to sulfide, giving rise to odour problems. It can also cause an increase in the concentration of ferrous iron in solution.

No health-based guideline value has been recommended for dissolved oxygen.

# 5. ACCEPTABILITY ASPECTS

### pH

Although pH usually has no direct impact on consumers, it is one of the most important operational water quality parameters. Careful attention to pH control is necessary at all stages of water treatment to ensure satisfactory water clarification and disinfection. For effective disinfection with chlorine, the pH should preferably be less than 8. The pH of the water entering the distribution system must be controlled to minimize the corrosion of water mains and pipes in household water systems (see section 6.6). Failure to do so can result in the contamination of drinking-water and in adverse effects on its taste, odour, and appearance.

The optimum pH required will vary in different supplies according to the composition of the water and the nature of the construction materials used in the distribution system, but it is often in the range 6.5–9.5. Extreme values of pH can result from accidental spills, treatment breakdowns, and insufficiently cured cement mortar pipe linings.

No health-based guideline value has been proposed for pH (see page 53).

### Sodium

The taste threshold concentration of sodium in water depends on the associated anion and the temperature of the solution. At room temperature, the average taste threshold for sodium is about 200 mg/litre.

As no firm conclusions can be drawn regarding the health effects of sodium, no health-based guideline value has been derived (see page 55).

### Sulfate

The presence of sulfate in drinking-water can cause noticeable taste. Taste impairment varies with the nature of the associated cation; taste thresholds have been found to range from 250 mg/litre for sodium sulfate to 1000 mg/litre for calcium sulfate. It is generally considered that taste impairment is minimal at levels below 250 mg/litre.

It has also been found that addition of calcium and magnesium sulfate (but not sodium sulfate) to distilled water improves the taste; optimal taste was recorded at 270 and 90 mg/litre for the two compounds, respectively.

As sulfate is one of the least toxic anions, no health-based guideline value has been derived (see page 55).

### Total dissolved solids

Total dissolved solids (TDS) can have an important effect on the taste of drinking-water. The palatability of water with a TDS level of less than 600 mg/litre is generally considered to be good; drinking-water becomes increasingly unpalatable at TDS levels greater than 1200 mg/litre. Water with extremely low concentrations of TDS may be unacceptable because of its flat, insipid taste.

The presence of high levels of TDS may also be objectionable to consumers owing to excessive scaling in water pipes, heaters, boilers, and household appliances. Water with concentrations of TDS below 1000 mg/litre is usually acceptable to consumers, although acceptability may vary according to local circumstances.

No health-based guideline value for TDS has been proposed (see page 56).

### Zinc

Zinc imparts an undesirable astringent taste to water. Tests indicate a taste threshold concentration of 4 mg/litre (as zinc sulfate). Water containing zinc at concentrations in excess of 5 mg/litre may appear opalescent and develop a greasy film on boiling, although these effects may also be noticeable at concentrations as low as 3 mg/litre. Although drinking-water seldom contains zinc at concentrations above 0.1 mg/litre, levels in tapwater can be considerably higher because of the zinc used in plumbing materials.

No health-based guideline value has been proposed for zinc in drinking-water (see page 57).

## 5.2.3 Organic constituents

### Toluene

Toluene has a sweet, pungent, benzene-like odour. The reported taste threshold ranges from 40 to 120 $\mu$g/litre. The reported odour threshold for toluene in water ranges from 24 to 170 $\mu$g/litre. Toluene may therefore affect the acceptability of water at concentrations below its health-based guideline value (see page 65).

### Xylenes

Xylene concentrations in the range 300–1000 $\mu$g/litre produce a detectable taste and odour. The odour threshold for xylene isomers in water has been reported to range from 20 to 1800 $\mu$g/litre. The lowest odour threshold is lower than the health-based guideline value derived for the compound (see page 65).

### Ethylbenzene

Ethylbenzene has an aromatic odour. The reported odour threshold for ethylbenzene in water ranges from 2 to 130 $\mu$g/litre. The lowest reported odour threshold is 100-fold lower than the health-based guideline value (see page 66). The taste threshold ranges from 72 to 200 $\mu$g/litre.

### Styrene

The average taste threshold reported for styrene in water at 40 °C is 120 $\mu$g/litre. Styrene has a sweet odour, and reported odour thresholds for styrene in water range from 4 to 2600 $\mu$g/litre, depending on temperature. Styrene may there-

# 5. ACCEPTABILITY ASPECTS

fore be detected in water at concentrations below its health-based guideline value (see page 66).

### Monochlorobenzene
Taste and odour thresholds of 10–20 µg/litre and odour thresholds ranging from 40 to 120 µg/litre have been reported for monochlorobenzene. The health-based guideline value derived for monochlorobenzene (see page 68) far exceeds the lowest reported taste and odour threshold in water.

### Dichlorobenzenes
Odour thresholds of 2–10 and 0.3–30 µg/litre have been reported for 1,2- and 1,4-dichlorobenzene, respectively. Taste thresholds of 1 and 6 µg/litre have been reported for 1,2- and 1,4-dichlorobenzene, respectively. The health-based guideline values derived for 1,2- and 1,4-dichlorobenzene (see page 69) far exceed the lowest reported taste and odour thresholds for these compounds.

### Trichlorobenzenes
Odour thresholds of 10, 5–30, and 50 µg/litre have been reported for 1,2,3-, 1,2,4-, and 1,3,5-trichlorobenzene, respectively. A taste and odour threshold concentration of 30 µg/litre has been reported for 1,2,4-trichlorobenzene. The health-based guideline value derived for total trichlorobenzenes (see page 70) exceeds the lowest reported odour threshold in water of 5 µg/litre.

### Synthetic detergents
In many countries, the earlier, persistent types of anionic detergent have been replaced by others that are more easily biodegraded, and hence the levels found in water sources have decreased substantially. New types of cationic, anionic, and non-ionic detergent have also been introduced. The concentration of detergents in drinking-water should not be allowed to reach levels giving rise to either foaming or taste or odour problems.

## 5.2.4 Disinfectants and disinfectant by-products

### Chlorine
The taste and odour thresholds for chlorine in distilled water are 5 and 2 mg/litre, respectively. Most individuals are able to taste chlorine or its by-products (e.g., chloramines) at concentrations below 5 mg/litre, and some at levels as low as 0.3 mg/litre, although a residual chlorine concentration of between 0.6 and 1.0 mg/litre will generally begin to cause problems with acceptability. The taste threshold of 5 mg/litre is at the health-based guideline concentration (see page 94).

Chlorophenols
Chlorophenols generally have very low organoleptic thresholds. The taste thresholds in water for 2-chlorophenol, 2,4-dichlorophenol, and 2,4,6-trichlorophenol are 0.1, 0.3 and 2 µg/litre, respectively. Odour thresholds are 10, 40, and 300 µg/litre, respectively. If water containing 2,4,6-trichlorophenol is free from taste, it is unlikely to present undue risk to health (see page 97).

# 6.
# Protection and improvement of water quality

## 6.1 General considerations

Compliance with drinking-water quality standards, based on these guidelines, should provide assurance that the supply is safe. However, it must be recognized that adequate monitoring is essential to ensure continuing compliance, and that there are many potential situations – some of which can arise very quickly – that could cause potentially hazardous situations to develop.

Many potential problems can be prevented by safeguarding the integrity of the raw water source and its watershed, by proper maintenance and inspection of the treatment plant and distribution system, by the training of managers and plant personnel, and by consumer education. However, although it is essential that water suppliers periodically reassess their operations to ensure that conditions that could affect the quality of water have not changed, that periodic maintenance is performed, that repairs and renewals of equipment are undertaken without delay when required, that personnel are adequately trained, and that job skills are maintained, a discussion of these important facets of water supply is outside the scope of this publication. The reader is referred to the many excellent texts available on these topics for guidance (see Bibliography).

Where piped water of high quality is continuously available to household connections, monitoring of the quality of this water provides an indication of the risk of waterborne diseases. Nevertheless, these conditions of water supply are, globally, the exception rather than the rule, and many people collect water from sources away from the point of use or store water in insanitary conditions in the household. Similarly, even with adequate conditions of supply, household storage tanks and domestic plumbing may be sources of contamination if not properly installed and maintained. For these reasons, water is subject to contamination in the household, and this may often be the most important source of microbiological contamination. Where household storage occurs, the surveillance agency should investigate the risk that this represents to human health, and remedial actions, such as education regarding water handling and promotion of maintenance of household storage tanks, should be instigated. This subject is considered further is Volume 3 of *Guidelines for drinking-water quality*.

It should be emphasized that, in terms of water quality, pathogenic microorganisms remain the most important danger to drinking-water in both developed and developing countries.

## 6.2 Selection and protection of water sources

Proper selection and protection of water sources are of prime importance in the provision of safe drinking-water. It is always better to protect water from contamination than to treat it after it has been contaminated.

Before a new source of drinking-water supply is selected, it is important to ensure that the quality of the water is satisfactory or treatable for drinking and that the quantity available is sufficient to meet continuing water demands, taking into account daily and seasonal variations and projected growth in the size of the community being served.

The watershed should be protected from human activities. This could include isolation of the watershed and/or control of polluting activities in the area, such as dumping of hazardous wastes, mining and quarrying, agricultural use of fertilizers and pesticides, and the limitation and regulation of recreational activities.

Sources of ground water such as springs and wells should be sited and constructed so as to be protected from surface drainage and flooding. Zones of ground water abstraction should be fenced to prevent public access, kept clean of rubbish, and sloped to prevent the collection of pools in wet weather. Animal husbandry should be controlled in such zones.

Protection of open surface water is a problem. It may be possible to protect a reservoir from major human activity, but, in the case of a river, protection may be possible only over a limited reach, if at all. Often it is necessary to accept existing and historical uses of a river or lake and to design the treatment accordingly.

## 6.3 Treatment processes

Water treatment processes used in any specific instance must take into account the quality and nature of the water supply source. The intensity of treatment must depend on the degree of contamination of the source water. For contaminated water sources, multiple treatment barriers to the spread of pathogenic organisms are particularly important and should be used to give a high degree of protection and to reduce the reliance on any individual treatment step.

The fundamental purpose of water treatment is to protect the consumer from pathogens and impurities in the water that may be offensive or injurious to human health. Urban treatment of water from lowland sources usually consists of

## 6. PROTECTION AND IMPROVEMENT OF WATER QUALITY

(1) reservoir storage or pre-disinfection, (2) coagulation, flocculation, and sedimentation (or flotation), (3) filtration, and (4) disinfection. Alternative or additional processes may be interposed to meet local conditions. Disinfection is the final safeguard and also protects drinking-water during distribution against external contamination and regrowth. The whole treatment sequence may indeed be regarded as conditioning the water for effective and reliable disinfection. Urban water treatment is, in effect, a four-stage multiple-barrier system for the removal of microbial contamination.

The multiple-barrier concept can be adapted for treating surface waters in rural and remote regions. A typical series of processes would include (1) storage, (2) sedimentation or screening, (3) gravel pre-filtration and slow-sand filtration, and (4) disinfection. Such treatment is considered in detail in Volume 3.

### 6.3.1 Pre-treatment

Surface waters may be either stored in reservoirs or disinfected before treatment.

During impoundment of water in lakes or reservoirs, the microbiological quality improves considerably as a result of sedimentation, the lethal effect of the ultraviolet content of sunlight in surface layers of water, and starvation and predation. Reductions of faecal indicator bacteria, salmonella, and enteroviruses are about 99%, being greatest during the summer and with residence periods of the order of 3–4 weeks.

Pre-disinfection is usual when water is abstracted and treated without storage. It will destroy animal life and reduce numbers of faecal bacteria and pathogens, besides assisting in the removal of algae during coagulation and filtration. An additional important function is the removal of ammonia. A drawback is that, when chlorine is used to excess, chlorinated organic compounds and biodegradable organic carbon will be produced.

Microstraining through very fine screens, typically with an average pore diameter of 30 $\mu$m, is an effective way of removing many microalgae and zooplankton that may otherwise clog or even penetrate filters. It has little, if any, effect in reducing numbers of faecal bacteria and enteric pathogens.

Where water of a very high quality is required, infiltration of raw or partly treated surface water into river banks or sand dunes can be practised, as notably in the Netherlands. Infiltration serves as a buffer in case raw river water cannot be used, because of incidents such as industrial pollution. The abstracted water usually needs additional treatment to remove iron or manganese compounds, and the detention period needs to be as long as possible to attain a quality approaching that of ground water. Removal of faecal bacteria and viruses exceeds 99%.

### 6.3.2 Coagulation, flocculation, and sedimentation

Coagulation involves the addition of chemicals (e.g., aluminium sulfate, ferrous or ferric sulfate, and ferric chloride) to neutralize the charges on particles and facilitate their agglomeration during the slow mixing provided in the flocculation step. Flocs thus formed co-precipitate, absorb, and entrap natural colour and mineral particles and can bring about major reductions in turbidity and in counts of protozoa, bacteria and viruses.

Coagulation and flocculation require a high level of supervisory skill. Before it is decided to use coagulation as part of a treatment process, careful consideration must be given to the likelihood of a regular supply of chemicals and the availability of qualified personnel.

The purpose of sedimentation is to permit settleable floc to be deposited and thus reduce the concentration of suspended solids that must be removed by filters. Among the factors that influence sedimentation are: size, shape, and weight of the floc; viscosity and hence temperature of the water; detention time; number, depth, and areas of the basins; surface overflow rate; velocity of flow; and inlet and outlet design. Plans must be made for the collection and safe disposal of sludge from sedimentation tanks. Flotation is an alternative to sedimentation when the amount of floc is slight.

For the coagulation/sedimentation process to be most effective for the control of trihalomethanes, the initial point of chlorine application should be after the coagulation/sedimentation process, to allow for as much precursor removal as possible prior to chlorination. Reductions in trihalomethane production of up to 75% in full-scale plants have been reported as a result of moving the initial chlorination application point past the coagulation/sedimentation process.

### 6.3.3 Rapid and slow sand filtration

When rapid filtration follows coagulation, its performance in removing microorganisms and turbidity varies through the duration of the run between backwashings. Immediately after backwashing, performance is poor, until the bed has compacted. Performance will also deteriorate progressively at the stage when backwashing is needed, as floc may escape through the bed into the treated water. These features emphasize the need for proper supervision and control of filtration at the waterworks.

Slow sand filtration is simpler to operate than rapid filtration, as frequent backwashing is not required. It is therefore particularly suitable for developing countries and small rural systems, but it is applicable only if sufficient land is available.

When the slow sand filter is first brought into use, a microbial slime community develops on the sand grains, particularly at the surface of the bed. This consists of bacteria, free-living ciliated protozoa and amoebae, crustacea, and invertebrate larvae acting in food chains, resulting in the oxidation of organic substances in the water and of ammoniacal nitrogen to nitrate. Pathogenic bacteria, viruses, and resting stages of parasites are removed, principally by adsorption and by subsequent predation. When correctly loaded, slow sand filtration brings about the greatest improvement in water quality of any single conventional water treatment process. Bacterial removal will be 98–99.5% or more, *E. coli* will be reduced by a factor of 1000, and virus removal will be even greater. A slow sand filter is also very efficient in removing parasites (helminths and protozoa). Slow sand filters are somewhat more effective when the water is warm. Nevertheless, the effluent from a slow sand filter might well contain a few *E. coli* and viruses, especially during the early phase of a filter run and with low water temperatures.

### 6.3.4 Disinfection

Terminal disinfection of piped drinking-water supplies is of paramount importance and is almost universal, as it is the final barrier to the transmission of waterborne bacterial and viral diseases. Although chlorine and hypochlorite are most often used, water may also be disinfected with chloramines, chlorine dioxide, ozone, and ultraviolet irradiation.

The efficacy of any disinfection process depends upon the water being treated beforehand to a high degree of purity, as disinfectants will be neutralized to a greater or lesser extent by organic matter and readily oxidizable compounds in water. Microorganisms that are aggregated or are adsorbed to particulate matter will also be partly protected from disinfection and there are many instances of disinfection failing to destroy waterborne pathogens and faecal bacteria when the turbidity was greater than 5 nephelometric turbidity units (NTU). It is therefore essential that the treatment processes preceding terminal disinfection are always operated to produce water with a median turbidity not exceeding 1 NTU and not exceeding 5 NTU in any single sample. Values well below these levels will regularly be attained with a properly managed plant.

Normal conditions of chlorination (i.e., a free residual chlorine of $\geqslant 0.5$ mg per litre, at least 30 minutes contact, pH less than 8.0, and water turbidity of less than 1 NTU) can bring about over 99% reductions of *E. coli* and certain viruses but not of the cysts or oocysts of parasitic protozoa.

The growth of bacteria within activated carbon point-of-use water filters has been well documented. Some manufacturers of carbon filters have attempted to avoid this problem by incorporating silver, as a bacteriostatic agent, in the filters. However, all of the published reports on this topic have convincingly demonstrated that this practice has a limited effect. It is believed that the presence of silver in these filters selectively permits the growth of silver-tolerant bacteria. For this reason, it is imperative that these devices be used only with drinking-water known to be microbiologically safe and that devices be well flushed prior to each use. Silver is occasionally used to disinfect drinking-water on board ships. However, because long contact times or high concentrations are essential, the use of silver for disinfection is not considered practical for point-of-use applications.

### 6.3.5 Fluoride removal

High fluoride levels, above 5 mg/litre, have been found in several countries (e.g., Algeria, China, Egypt, India, and Thailand). Such high levels have at times led to dental or skeletal fluorosis.

Fluoride removal techniques have been developed for both community water supplies and individual households. The most frequently employed fluoride removal technique uses ion exchange/adsorption with either charred bone-meal or activated alumina. Full-scale activated alumina facilities and household defluoridators using charred bone-meal have been reported to reduce fluoride levels from 5–8 mg/litre to less than 1 mg/litre. Fluoride-spent bone-meal and activated alumina are usually regenerated for further use.

## 6.4 Choice of treatment

In small communities in rural areas, protection of the source of water may be the only form of treatment possible. Such supplies are considered in detail in Volume 3. Where communities are large, the demand for water is high and can often be met only by using additional sources of poor microbiological quality. Such waters will require all the resources of water treatment to yield an attractive and safe drinking-water.

Ground waters extracted from deep, well-protected aquifers are usually free from pathogenic microorganisms, and the distribution of such untreated ground water is common practice in many countries. This practice implies that the area of influence is protected by effective regulatory measures and that the distribution system is adequately protected against secondary contamination of the drinking-water. If continuous protection from source to consumer cannot be guaranteed, then disinfection and the maintenance of adequate concentrations of residual chlorine are imperative.

Surface water will usually require full treatment. The degrees of removal of microorganisms by coagulation, flocculation, sedimentation, and rapid filtration are, with proper design and operation, equivalent to those for slow sand filtration.

Additional treatment, such as ozonation, followed by activated carbon treatment to remove assimilable organic carbon, reduces the potential for aftergrowth problems caused by nuisance bacteria in distribution networks. The ozonation stage may also have a significant effect on reducing pathogens. Disinfection should be regarded as obligatory for all piped supplies using surface water, even those derived from high-quality, unpolluted sources, as there should always be more than one barrier against the transmission of infection in a water supply. In large, properly run waterworks, the criteria for the absence of *E. coli* and coliform bacteria can then be met with a very high degree of probability. The current trend is to optimize the use of chemicals such as chlorine and coagulants in water treatment, and to develop physical or biological methods of treatment, in order to reduce the doses of chemicals required, thereby reducing the formation of disinfection by-products.

## 6.5 Distribution networks

The distribution network transports water from the place of treatment to the consumer. Its design and size will be governed by the topography and the location and size of the community. The aim should always be to ensure that consumers receive a sufficient and uninterrupted supply, and that contamination is not introduced in transit.

Distribution systems are especially vulnerable to contamination when the pressure falls, particularly in the intermittent supplies of many cities in developing countries. Suction is often created by direct pumping from the mains to private storage tanks, a practice that should be minimized.

The bacteriological quality of water can deteriorate during distribution. If the water contains significant assimilable organic carbon or ammonia, adequate residual levels of disinfectant are not maintained. If such water-mains are not flushed and cleaned frequently enough, growth of nuisance bacteria and other organisms can occur. Where the water contains appreciable assimilable organic carbon (>0.25 mg/litre) and where the water temperature exceeds 20 °C, a concentration of residual free chlorine of 0.25 mg/litre may be required to prevent growth of *Aeromonas* and other nuisance bacteria. Attached microorganisms may grow even in the presence of residual chlorine. The aim should be to produce biologically stable water, with very low levels of organic compounds and ammonia to prevent problems from microbial growth in distribution.

Underground storage tanks and service reservoirs must be inspected for deterioration and for infiltration of surface and ground water. It is desirable for the land enclosing underground storage tanks to be fenced off to prevent access by humans and animals and to prevent damage to the structures.

Repair works to mains offer another possibility for contamination. Local loss of pressure may result in back-siphonage of contaminated water, unless check valves are introduced into the water system at sensitive points, such as supplies to garden irrigation and urinals. If the main has been damaged and if there is the possibility that wastewater from a fractured sewer or drain may have entered, the situation is most serious. The actions that must be taken to protect consumers from waterborne disease should be specified in national codes of practice and in local instructions to waterworks staff.

Microbial contamination can occur by growth on unsatisfactory construction materials coming into contact with water, such as washers, pipe lining compounds, and plastics used in pipes and taps. National systems should be in operation controlling the use of such materials.

## 6.6 Corrosion control

### 6.6.1 Introduction

Corrosion is characterized by the partial solubilization of the materials constituting the treatment and supply systems, tanks, pipes, valves, and pumps. It may lead to structural failure, leaks, loss of capacity, and deterioration of chemical and microbiological water quality. The internal corrosion of pipes and fittings can have a direct impact on the concentration of some water constituents for which guideline values have been recommended, including cadmium, copper, iron, lead, and zinc. Corrosion control is therefore an important aspect of the management of a water supply system.

Because of its implications for water quality, the present discussion will deal only with the internal corrosion of pipes; the protection of pipes against external corrosion is extremely important, but is much less relevant to water quality.

Corrosion control involves many parameters, including the concentrations of calcium, bicarbonate, carbonate, and dissolved oxygen, as well as pH. The detailed requirements differ for every water and for each distribution material.

### 6.6.2 Basic considerations

Many metals, including most of those used in the construction of water supply systems, are unstable in the presence of water and have a tendency to transform or degrade to a more stable and often soluble form – a process recognizable as corrosion. The rate at which this takes place is governed by many chemical and physical factors; it may be very rapid or extremely slow.

Of great importance are the properties of the products of corrosion, the stable end-products of the process. If any of these is soluble in water, then corrosion will tend to be rapid. In some cases, however, where the corrosion products are insoluble, a protective scale may be formed at the water surface, and corrosion then becomes very slow. Insoluble corrosion products are protective only where they form an impenetrable layer. If they form a spongy or flocculent mass, corrosion will continue, leading to a deterioration of water quality, a reduction of the carrying capacity of the pipe, and microbial growths (biofilms), which may be protected from residual chlorine.

Corrosion is also greatly influenced by the electrical properties of the metals in contact with water. Different metals show different tendencies to develop an electric charge in contact with water, and this difference is displayed in the so-called galvanic series. Where two different metals (or other electrically conducting materials) are in contact, a galvanic cell is formed in which metal will dissolve at the negative electrode. It is not necessary for the two metals involved to be at the same location provided that they are in electrical contact. The formation of galvanic cells often provides the driving force for corrosion.

The rate of corrosion is governed mainly by the rate at which dissolved reactants are transported to the metal surface and the rate at which dissolved products are transported away from the reaction site. Thus, corrosion rates increase directly with increasing concentration of ions in the water and also with increasing degrees of agitation.

At very high water velocities, the rate of corrosion may increase dramatically as a result of erosion corrosion. In common with other chemical reactions, corrosion rates increase with temperature.

Certain metals undergo a phenomenon known as passivation. For these metals, which include iron, nickel, and chromium, and their alloys, the application of a certain voltage results in a substantial decrease in corrosion rate, which is maintained over a considerable range of applied voltage. The process is exploited in some corrosion control strategies, including "anodic protection". Copper, lead, and zinc corrosion cannot be controlled by anodic protection.

### 6.6.3 Effect of water composition

Dissolved oxygen is one of the most important factors influencing the rate of corrosion. It is a direct participant in the corrosion reaction, and, under most circumstances, the higher its concentration the higher the corrosion rate.

pH controls the solubility, rate of reaction, and, to some extent, the surface chemistry of most of the metal species involved in corrosion reactions. It is particularly important in relation to the formation of a protective film at the metal surface.

There is increasing evidence of the importance of the aggressive action of the chloride ion in the corrosion of metals used in distribution systems. There is some evidence that residual chlorine also affects the rate of corrosion.

### 6.6.4 Corrosion of pipe materials

#### *Copper*

Copper tubing may be subject to general corrosion, impingement attack, and pitting corrosion. General corrosion of copper is most often associated with soft, acidic waters; waters with a pH below 6.5 and a hardness of less than 60 mg/litre (as $CaCO_3$) are very aggressive to copper and should not be transported in copper pipes or heated in copper boilers. Impingement attack is the result of excessive flow velocities and is aggravated in soft water at high temperature and low pH. The pitting of copper is commonly associated with hard ground waters having a carbon dioxide concentration above 5 mg/litre and a high dissolved oxygen level. Surface waters containing organic colour (humic substances) may also be associated with pitting corrosion. A high proportion of general and pitting corrosion problems are associated with new pipes in which a protective oxide layer has not yet formed.

#### *Lead*

The corrosion of lead (plumbosolvency) is of particular concern because of its adverse effect on water quality. Lead piping is still common in old houses, and lead solders have been used widely, particularly for jointing copper tube. Lead is stable in water in a number of forms, depending on pH, and the solubility of lead is governed to a large extent by the formation of insoluble lead carbonate. The solubility of lead increases markedly as the pH is reduced below 8 because of the substantial decrease in the equilibrium carbonate concentration. Thus, plumbosolvency tends to be at a maximum in waters with a low pH and low alkalinity, and a useful interim control procedure pending pipe replacement is to maintain pH in the range 8.0–8.5.

#### *Cement and concrete*

Concrete is a composite material consisting of a cement binder in which an inert aggregate is embedded. Cement is primarily a mixture of calcium silicates and aluminates together with some free lime. Cement mortar, in which the aggregate is fine sand, is used as a protective lining in iron and steel water pipes. In asbestos–cement (A/C) pipes, the aggregate is asbestos fibres. Cement is subject to deterioration on prolonged exposure to aggressive water – due either to the dissolution of lime and other soluble compounds or to chemical attack by aggressive ions such as chloride or sulfate – and this may result in structural failure of the cement pipe. The aggressiveness of a water to cement is related to the value of the Langelier index, which measures the potential for precipitation or dissolution of calcium carbonate (see section 6.6.6). There is also a similar "aggressivity index", which has been used specifically to assess the potential for the dissolution of concrete. A pH of 8.5 or higher may be necessary to control cement corrosion.

### 6.6.5 Microbiological aspects of corrosion

Microorganisms can play a significant role in the corrosion of pipe material by forming micro-zones of low pH or high concentrations of corrosive ions, mediating oxidation processes or the removal of corrosion products, and disrupting protective surface films. The most significant bacteria involved in corrosion are the sulfate-reducing and the iron bacteria, but nitrate reducers and methane producers may have a role in some situations. Corrosion induced by microorganisms tends to be a problem in distribution systems where a residual concentration of chlorine has not been maintained, especially in "dead ends" and other situations where the flow is low. It may also be a problem where there has been heavy scale deposition or where bulky corrosion products have formed.

### 6.6.6 Corrosion indices

A number of indices have been developed to characterize the corrosion potential of any particular water. Most are based on the assumption that water with a tendency to deposit a calcium carbonate scale on metal surfaces will be less corrosive. Thus, the well-known Langelier index is the difference between the actual pH of a water and its "saturation pH", this being the pH at which a water of the same alkalinity and calcium hardness would be at equilibrium with solid calcium carbonate. In addition to the calcium hardness and alkalinity, the calculation of the saturation pH takes account of the concentration of the total dissolved solids and the temperature.

Waters with a pH higher than their saturation pH (positive Langelier index) are supersaturated with respect to calcium carbonate and will therefore tend to deposit a scale. Conversely, waters with a pH lower than their saturation pH (negative Langelier index) will be undersaturated with respect to calcium carbonate and are therefore considered to be aggressive. Nomographs are available to simplify the determination of the saturation pH. Ideally, distributed water should be at or slightly above its saturation pH.

The Langelier index and other indices based on similar principles have proved to be helpful in predicting and dealing with corrosion problems in many situations. Clearly, however, the assumption that a calcium carbonate scale will always be protective and that water that does not lay down such a scale will always be corrosive oversimplifies a complex phenomenon. It is not surprising, therefore, that attempts to quantify aggressiveness on this basis have produced mixed results.

The ratio of the chloride and sulfate concentrations to the bicarbonate concentration (Larson ratio) has been shown to be helpful in assessing the corrosiveness of water to cast iron and steel. A similar approach has been used in studying dissolution of zinc from brass fittings.

### 6.6.7 Strategies for corrosion control

The main strategies for corrosion control include:

— the control of environmental parameters affecting corrosion rate,
— the addition of chemical inhibitors,
— electrochemical measures, and
— considerations of system design.

To control corrosion in water distribution networks, the methods most commonly applied are controlling pH, increasing the carbonate hardness, or adding corrosion inhibitors such as sodium polyphosphates or silicates and zinc orthophosphate. The quality and maximum dose to be used should be in line with appropriate national specifications for such water treatment chemicals. Although pH control is an important approach, its possible impact on other aspects of water supply technology, including disinfection, must always be taken into account.

## 6.7 Emergency measures

It is essential that water suppliers develop contingency plans to be invoked in the event of an emergency. These plans should consider potential natural disasters (such as earthquakes, floods, damage to electrical equipment by lightning strikes), accidents (spills in the watershed), damage to treatment plant and distribution system, and human actions (strikes, sabotage). Contingency plans should clearly specify responsibilities for coordinating measures to be taken, a communication plan to alert and inform users of the supply, and plans for providing and distributing emergency supplies of water.

In an emergency, a decision to close the supply carries an obligation to provide an alternative safe supply. Advising consumers to boil water, initiating superchlorination, and undertaking immediate corrective measures may be preferable. National drinking-water standards are intended to ensure that the consumer enjoys safe potable water, not to shut down deficient water supplies.

During an emergency in which there is evidence of faecal contamination of the supply, it may be necessary either to modify the treatment of existing sources or temporarily to use alternative sources of water. It may be necessary to increase disinfection at source or to rechlorinate during distribution. If possible, the distribution system should be kept under continuous pressure, as failure in this respect will considerably increase the risks of entry of contamination to the pipework and thus the possibility of waterborne disease. If the quality cannot be maintained, consumers should be advised to boil the water during the emergency. The water should be brought to a vigorous rolling boil for 1 minute. As water boils at a lower temperature at high altitude, a minute of extra boiling time should be added for every 1000 m above sea-level. This should kill or inactivate the vegetative cells of bacteria and viruses as well as the cysts of *Giardia*. If bulk supplies in tankers are used, sufficient chlorine should be added to ensure that a free residual concentration of at least 0.5 mg/litre for a minimum of 30 minutes is present at the delivery point. Before use, tankers should be either disinfected or steam-cleaned. The temporary use of other disinfectant measures, such as slow-release disinfectant tablets added to water drawn from the tap, should also be considered if they have been proven to give safe and reliable disinfection.

It is impossible to give general guidance concerning emergencies in which chemicals cause massive contamination of the supply. The guideline values recommended relate to a level of exposure that is regarded as tolerable throughout life; acute toxic effects are not normally considered in the assessment of a TDI. The length of time during which exposure to a chemical far in excess of the guideline value would be toxicologically detrimental will depend upon factors that vary from contaminant to contaminant. The biological half-life of the contaminant, the nature of the toxicity, and the amount by which the exposure exceeds the guideline value are all crucial. In an emergency situation the public health authorities must be consulted about appropriate action.

# Bibliography

## Chapter 2. Microbiological aspects

Pathogenic agents and control of waterborne disease

Falconer IR, Beresford AM, Runnegar MTC. Evidence of liver damage by toxin from a bloom of the blue-green algae, *Microcystis aeruginosa. Medical journal of Australia,* 1983, 1: 511-514.

Galbraith NS et al. Water and disease after Croydon: a review of water-borne and water-associated diseases in the UK 1937-1986. *Journal of the Institution of Water and Environmental Management,* 1987, 1: 7-21.

Lippy EC, Waltrip SC. Waterborne disease outbreaks – 1946–1980: a thirty-five year perspective. *Journal of the American Water Works Association,* 1984, 76(2): 60-67.

Regli S et al. Modelling the risk from *Giardia* and viruses in drinking water. *Journal of the American Water Works Association,* 1991, 83(11): 76-84.

Short CS. The Bramham incident, 1980 – an outbreak of water-borne infection. *Journal of the Institution of Water and Environmental Management,* 1988, 2: 383-390.

Steering Committee for Cooperative Action for the International Drinking Water Supply and Sanitation Decade. *Report on IDWSSD impact on diarrheal disease.* Geneva, World Health Organization, 1990.[1]

Steering Committee for Cooperative Action for the International Drinking Water Supply and Sanitation Decade. *Report on IDWSSD impact on dracunculiasis.* Geneva, World Health Organization, 1990.[1]

Steering Committee for Cooperative Action for the International Drinking Water Supply and Sanitation Decade. *Report on IDWSSD impact on schistosomiasis.* Geneva, World Health Organization, 1990.[1]

World Health Organization. *Surveillance of drinking-water quality.* Geneva, 1976 (Monograph Series, No. 63).

---

[1] Unpublished document, available from Community Water Supply and Sanitation, World Health Organization, 1211 Geneva 27, Switzerland.

## Standard microbiological methods

American Public Health Association. *Standard methods for the examination of water and wastewater*, 17th ed. Washington, DC, 1989.

Block J-C, Schwartzbrod L. *Analyse virologique des eaux. Techniques de mise en évidence de virus humains.* Paris, Technique et Documentation, Lavoisier, 1982.

Codex Alimentarius Commission. *Codex standards for natural mineral waters and edible ices and ice mixes.* Rome, Food and Agriculture Organization of the United Nations, Codex Alimentarius Vol. XII, 1st ed., 1982, and Suppl. 1, 1986.

Department of Health and Social Security. *The bacteriological examination of drinking water supplies 1982.* London, Her Majesty's Stationery Office, 1983 (Reports on Public Health and Medical Subjects No. 71).

Maul A, Vagost D, Block J-C. *Stratégie d'échantillonnage pour l'analyse microbiologique sur les réseaux de distribution d'eau.* Paris, Lavoisier, 1989.

# Chapter 3. Chemical aspects

## Sampling and analytical methods

American Public Health Association. *Standard methods for the examination of water and wastewater*, 17th ed. Washington, DC, 1989.

International Organization for Standardization. *Water quality series.* Geneva.

Rodier J. *L'analyse de l'eau. Eaux naturelles, eaux résiduaires, eau de mer.* 7th ed. Paris, Dunod, 1984.

## Risk assessment

Bull RJ, Kopfler FC. *Health effects of disinfectants and disinfection by-products.* Denver, CO, American Waterworks Association, 1991.

Environmental Health Criteria Series. Geneva, World Health Organization.
*Pentachlorophenol* (No. 71, 1987).
*Permethrin* (No. 94, 1990).
*Methylmercury* (No. 101, 1990).
*Beryllium* (No. 106, 1990).
*Barium* (No. 107, 1990).
*Nickel* (No. 108, 1990).
*Tributyltin compounds* (No. 116, 1990).
*Inorganic mercury* (No. 118, 1990).
*Aldicarb* (No. 121, 1991).
*Lindane* (No. 124, 1991).
*Chlorobenzenes other than hexachlorobenzene* (No. 128, 1991).

*Diethylhexylphthalate* (No. 131, 1992).
*Cadmium* (No. 134, 1992).
*1,1,1-Trichloroethane* (No. 136, 1992).

International Agency for Research on Cancer. *Overall evaluations of carcinogenicity: an updating of IARC Monographs volumes 1 to 42.* Lyon, 1987 (IARC Monographs on the Evaluation of Carcinogenic Risks to Humans, Suppl. 7).

International Agency for Research on Cancer. *Chlorinated drinking-water; chlorination by-products; some other halogenated compounds; cobalt and cobalt compounds.* Lyon, 1991 (IARC Monographs on the Evaluation of Carcinogenic Risks to Humans, Volume 52).

Joint FAO/WHO Expert Committee on Food Additives. *Evaluation of certain food additives and the contaminants mercury, lead, and cadmium:* sixteenth report. Geneva, World Health Organization, 1972 (WHO Technical Report Series, No. 505).

Joint FAO/WHO Expert Committee on Food Additives. *Evaluation of certain food additives and contaminants.* Geneva, World Health Organization.
Twenty-second report, 1978 (WHO Technical Report Series, No. 631).
Twenty-sixth report, 1982 (WHO Technical Report Series, No. 683).
Twenty-seventh report, 1983 (WHO Technical Report Series, No. 696).
Twenty-eighth report, 1984 (WHO Technical Report Series, No. 710).
Thirtieth report, 1987 (WHO Technical Report Series, No. 751).
Thirty-third report, 1989 (WHO Technical Report Series, No. 776).
Thirty-seventh report, 1991 (WHO Technical Report Series, No. 806).

International Programme on Chemical Safety. *Summary of toxicological evaluations performed by the Joint FAO/WHO Meeting on Pesticide Residues (JMPR).* Geneva, World Health Organization, 1991 (unpublished document, WHO/PCS/92.9; available from Programme for the Promotion of Chemical Safety, World Health Organization, 1211 Geneva 27, Switzerland).

National Research Council. *Drinking water and health*, Vol. 1, 1977, to Vol. 9, 1989. Washington, DC, National Academy Press.

National Research Council. *Recommended dietary allowances,* 10th ed. Washington, DC, National Academy Press, 1989.

## Chapter 4. Radiological aspects

American Public Health Association, *Standard methods for the examination of water and wastewater,* 17th ed. Washington, DC, 1989.

Optimization and decision-making in radiological protection. *Annals of the ICRP,* 1989, 20 (1).

1990 Recommendations of the International Commission on Radiological Protection. *Annals of the ICRP,* 1990, 21 (1–3).

Association of Official Analytical Chemists. *Official methods of analysis of the Association of Official Analytical Chemists,* 15th ed. Washington, DC, 1990.

Environmental Measurements Laboratory. *EML procedures manual.* New York, Department of Energy, 1990 (HASL-300).

International Organization for Standardization. *Water quality – measurement of gross alpha activity in non-saline water – thick source method.* Geneva, 1990 (Draft International Standard 9696).

International Organization for Standardization. *Water quality – measurement of gross beta activity in non-saline water.* Geneva, 1990 (Draft International Standard 9697).

National Council on Radiation Protection and Measurements. *Control of radon in houses. Recommendations of the National Council on Radiation Protection and Measurements.* Bethesda, MD, 1989 (NCRP Report No. 103).

National Radiological Protection Board. *Committed equivalent organ doses and committed effective doses from intakes of radionuclides.* A report of the National Radiological Protection Board of the United Kingdom. Chilton, Didcot, 1991 (NRPB–R245).

Suess MJ, ed. *Examination of water for pollution control.* 3 vols. Oxford, Pergamon Press, 1982.

United States Environmental Protection Agency. Eastern Environmental Radiation Facility. *Radiochemistry procedures manual.* Montgomery, AL, 1987 (EPA 520/5-84-006).

United Nations Scientific Committee on the Effects of Atomic Radiation. *Sources, effects and risks of ionizing radiation.* New York, United Nations, 1988.

World Health Organization. *Derived intervention levels for radionuclides in food.* Geneva, 1988.

## Chapter 5. Acceptability aspects

Department of National Health and Welfare (Canada). *Guidelines for Canadian drinking water quality. Supporting documentation.* Ottawa, 1980.

National Institute for Water Supply. *Compilation of odour threshold values in air and water.* Zeist, Netherlands, 1977.

Zoetman BCJ. *Sensory assessment of water quality.* New York, Pergamon Press, 1980.

## Chapter 6. Protection and improvement of water quality

Abram FSH et al. *Permethrin for the control of animals in water mains.* Medmenham, Water Research Centre, 1980 (Technical Report No. 145).

American Water Works Association. *Water quality and treatment.* 4th ed. New York, McGraw-Hill, 1990.

Cox CR. *Operation and control of water treatment processes.* Geneva, World Health Organization, 1969 (Monograph Series, No. 49).

Degrémont. *Water treatment handbook,* 6th ed. Paris, Lavoisier, 1991.

Department of the Environment, Welsh Office. *Guidance on safeguarding the quality of public water supplies.* London, Her Majesty's Stationery Office, 1989.

Department of National Health and Welfare (Canada). *Guidelines for Canadian drinking water quality. Application manual for the production of drinking water.* Ottawa, Canadian Government Publishing Centre (in press).

Dupont A. *Hydraulique urbaine. Tome 1: Hydrologie, captage et traitement des eaux.* 1986. *Tome 2: Ouvrages de transport. Elévation et distribution des eaux,* 1988. Paris, Eyrolles.

Lallemand-Barres A, Roux J-C. *Guide méthodologique d'établissement des périmètres de protection des captages d'eau souterraine destinée à la consommation humaine.* Orléans, Editions du BRGM, 1986 (Coll. Manuels et Méthodes, No. 19).

Montout G, Larguier M. *Protection des distributions d'eau.* Paris, Compagnie générale des Eaux, Laboratoire d'hygiène de la ville de Paris, 1979.

Rajagopalan S, Shiffman MA. *Guide to simple sanitary measures for the control of enteric diseases.* Geneva, World Health Organization, 1974.

Water Authorities Association. *Guide to the microbiological implications of emergencies in the water services.* London, 1985.

World Health Organization. *Surveillance of drinking-water quality.* Geneva, 1976 (Monograph Series, No. 63).

WHO Regional Office for Europe. *Disinfection of rural and small-community water supplies.* Medmenham, Water Research Centre, 1989.

Annex 1
# List of participants in preparatory meetings

Consultation on Revision of WHO Guidelines for Drinking-Water Quality (Rome, Italy, 17–19 October 1988)

## Members

L. Albanus, Head, Toxicology Laboratory, National Food Administration, Uppsala, Sweden

J. Alexander, Toxicological Department, National Institute of Public Health, Oslo, Norway

J.A. Cotruvo, Director, Criteria and Standards Division, United States Environmental Protection Agency, Washington, DC, USA

H. de Kruijf, Laboratory for Ecotoxicology, Environmental Chemistry and Drinking-Water, National Institute of Public Health and Environmental Protection, Bilthoven, Netherlands

H.H. Dieter, Director and Professor, Institute for Water, Soil and Air Hygiene of the Federal Office of Health, Berlin

J.K. Fawell, Principal Toxicologist, Water Research Centre, Medmenham, England (*Rapporteur*)

E. Funari, Department of Environmental Hygiene, Istituto Superiore di Sanità, Rome, Italy

J.R. Hickman, Acting Director-General, Environmental Health Directorate, Health and Welfare Canada, Ottawa, Canada

Y. Magara, Director, Department of Sanitary Engineering, Institute of Public Health, Tokyo, Japan

R.F. Packham, Chief Scientist, Water Research Centre, Medmenham, England

M. Waring, Department of Health and Social Security, London, England

G.A. Zapponi, Environmental Impact Assessment Section, Istituto Superiore di Sanità, Rome, Italy

## Observers

S. Blease, Administrator, Water Protection Division, Commission of European Communities, Brussels, Belgium

B. Julin, Regulatory Affairs Manager, International Group of National Associations of Manufacturers of Agrochemical Products, Wilmington, DE, USA

A. Pelfrène, International Group of National Associations of Manufacturers of Agrochemical Products, Paris, France

N. Sarti, Division of Water and Soil, Ministry of Health, Rome, Italy

*Secretariat*

G. Burin, International Programme on Chemical Safety, Division of Environmental Health, World Health Organization, Geneva, Switzerland

R. Helmer, Prevention of Environmental Pollution, Division of Environmental Health, World Health Organization, Geneva, Switzerland

M. Mercier, Manager, International Programme on Chemical Safety, Division of Environmental Health, World Health Organization, Geneva, Switzerland

G. Ozolins, Manager, Prevention of Environmental Pollution, Division of Environmental Health, World Health Organization, Geneva, Switzerland (*Moderator*)

S. Tarkowski, Director, Environment and Health, WHO Regional Office for Europe, Copenhagen, Denmark

## Microbiology Consultation (London, England, 23 June 1989)

*Members*

U. Blumental, London School of Hygiene and Tropical Medicine, London, England

S. Cairncross, London School of Hygiene and Tropical Medicine, London, England

A.H. Havelaar, National Institute of Public Health and Environmental Protection, Bilthoven, Netherlands

R.F. Packham, Marlow, England

W. Stelzer, Research Institute of Hygiene and Microbiology, Bad Elster, German Democratic Republic

H. Utkilen, Department of Sanitary Engineering and Environmental Protection, National Institute of Public Health, Oslo, Norway

R. Walter, Director, Institute for General and Community Hygiene, Dresden, German Democratic Republic

*Secretariat*

J.K. Fawell, Principal Toxicologist, Water Research Centre, Medmenham, England

R. Helmer, Prevention of Environmental Pollution, Division of Environmental Health, World Health Organization, Geneva, Switzerland

B. Lloyd, Environmental Health Unit, Robens Institute of Industrial and Environmental Health and Safety, Guildford, England

E.B. Pike, Water Research Centre, Medmenham, England

## Coordination Consultation (Copenhagen, Denmark, 4–5 September 1989)

### *Members*

J.K. Fawell, Principal Toxicologist, Water Research Centre, Medmenham, England (*Co-Rapporteur*)

E. Funari, Department of Environmental Hygiene, Istituto Superiore di Sanità, Rome, Italy

E.S. Jensen, Senior Technical Adviser on Water Supply and Sanitation Projects, Technical Advisory Division, Danish International Development Agency, Copenhagen, Denmark

A. Minderhoud, Laboratory for Ecotoxicology, Environmental Chemistry and Drinking Water, National Institute of Public Health and Environmental Protection, Bilthoven, Netherlands

B. Mintz, Chief, Health Effects Assessment Section, Criteria and Standards Division, Office of Drinking-Water, United States Environmental Protection Agency, Washington, DC, USA

P.A. Nielsen, Scientific Officer, Toxicologist, Institute of Toxicology, National Food Agency, Soborg, Denmark

E. Poulsen, Chief Adviser in Toxicology, Institute of Toxicology, National Food Agency, Soborg, Denmark

B. Schultz, Water Quality Institute, Horsholm, Denmark

### *Secretariat*

G. Burin, International Programme on Chemical Safety, Division of Environmental Health, World Health Organization, Geneva, Switzerland (*Co-Rapporteur*)

O. Espinoza, Regional Officer for International Water Decade, Environment and Health, WHO Regional Office for Europe, Copenhagen, Denmark

R. Helmer, Prevention of Environmental Pollution, Division of Environmental Health, World Health Organization, Geneva, Switzerland

D. Kello, Project Officer for Toxicology and Food Safety, Environment and Health, WHO Regional Office for Europe, Copenhagen, Denmark

S. Tarkowski, Director, Environment and Health, WHO Regional Office for Europe, Copenhagen, Denmark

## Coordination Group Meeting (Geneva, Switzerland, 13–14 March 1990)

### Members

J.K. Fawell, Principal Toxicologist, Water Research Centre, Medmenham, England

E. Funari, Department of Environmental Hygiene, Istituto Superiore di Sanità, Rome, Italy

J.R. Hickman, Acting Director-General, Environmental Health Directorate, Health and Welfare Canada, Ottawa, Canada

A. Minderhoud, Laboratory for Ecotoxicology, Environmental Chemistry and Drinking Water, National Institute of Public Health and Environmental Protection, Bilthoven, Netherlands

B. Mintz, Chief, Health Effects Assessment Section, Criteria and Standards Division, Office of Drinking-Water, United States Environmental Protection Agency, Washington, DC, USA

B. Schultz, Water Quality Institute, Horsholm, Denmark

### Secretariat

G. Burin, International Programme on Chemical Safety, Division of Environmental Health, World Health Organization, Geneva, Switzerland

O. Espinoza, Regional Officer for International Water Decade, Environment and Health, WHO Regional Office for Europe, Copenhagen, Denmark

R. Helmer, Prevention of Environmental Pollution, Division of Environmental Health, World Health Organization, Geneva, Switzerland

D. Kello, Project Officer for Toxicology and Food Safety, Environment and Health, WHO Regional Office for Europe, Copenhagen, Denmark

G. Ozolins, Manager, Prevention of Environmental Pollution, Division of Environmental Health, World Health Organization, Geneva, Switzerland

R. Plestina, International Programme on Chemical Safety, Division of Environmental Health, World Health Organization, Geneva, Switzerland

## First Review Group Meeting on Pesticides (Busto Garolfo, Italy, 25–30 June 1990)

### Members

H. Abouzaid, Chief, Water Quality Control Division, National Agency for Drinking-Water, Rabat-Chellah, Morocco

# ANNEX 1

H. Atta-ur-Rahman, Director, H.E.J. Research Institute of Chemistry, Karachi, Pakistan

V. Benes, Chief, Toxicology and Reference Laboratory, Institute of Hygiene and Epidemiology, Prague, Czechoslovakia

J.F. Borzelleca, Pharmacology, Toxicology, Medical College of Virginia, Virginia Commonwealth University, Richmond, VA, USA

L. Brener, Chief, Department of Mineral Analysis, Research Laboratory, Société Lyonnaise des Eaux-Dumez, Paris, France

D. Calamari, Institute of Agricultural Entomology, Faculty of Agriculture, University of Milan, Italy

J. Du, Office of Drinking-Water, United States Environmental Protection Agency, Washington, DC, USA

J. K. Fawell, Principal Toxicologist, Water Research Centre, Medmenham, England (*Rapporteur*)

J. Forslund, National Agency of Environmental Protection, Copenhagen, Denmark

E. Funari, Department of Environmental Hygiene, Istituto Superiore di Sanità, Rome, Italy

A. Jaron, Commission of the European Communities, Brussels, Belgium

M. Maroni, Director, International Centre for Pesticide Safety, Busto Garolfo, Italy

Y. Patel, Health Effects Assessment, Office of Drinking-Water, United States Environmental Protection Agency, Washington, DC, USA

E. Poulsen, Chief Adviser in Toxicology, Institute of Toxicology, National Food Agency, Soborg, Denmark (*Chairman*)

J. Rueff, Department of Genetics, Faculty of Medical Science, Lisbon, Portugal

B. Schultz, Water Quality Institute, Horsholm, Denmark

J.A. Sokal, Head, Department of Toxicity Evaluation, Institute of Occupational Medicine, Lodz, Poland

M. Takeda, Director of Environmental Chemistry, National Institute of Hygienic Science, Tokyo, Japan

E.M. den Tonkelaar, National Institute of Public Health and Environmental Protection, Bilthoven, Netherlands

G. Wood, Acting Head, Criteria Section, Monitoring and Criteria Division, Environmental Health Directorate, Health and Welfare, Ottawa, Canada

## *Observers*

S. Behrendt, BASF AG, Limburgerhof, Federal Republic of Germany

S. Hahn, BASF AG, Limburgerhof, Federal Republic of Germany

H. Kieczka, BASF AG, Limburgerhof, Federal Republic of Germany

S. Kimura, Southern Fukuoka Prefecture, Water Spread Authority, Japan Water Works Association, Tokyo, Japan

F. Sarhan, CIBA-GEIGY Ltd, Basel, Switzerland

G.E. Veenstra, Shell International Petroleum, The Hague, Netherlands

### Secretariat

G. Burin, International Programme on Chemical Safety, Division of Environmental Health, World Health Organization, Geneva, Switzerland

D. Kello, Project Officer for Toxicology and Food Safety, Environment and Health, WHO Regional Office for Europe, Copenhagen, Denmark

R. Plestina, International Programme on Chemical Safety, Division of Environmental Health, World Health Organization, Geneva, Switzerland

## First Review Group Meeting on Organics (Copenhagen, Denmark, 6–10 November 1990)

### Members

C. Abernathy, Toxicologist, Health Effects Branch, Office of Drinking-Water, United States Environmental Protection Agency, Washingon, DC, USA

H.H. Dieter, Director and Professor, Institute for Water, Soil and Air Hygiene of the Federal Office of Health, Berlin, Germany

A.M. van Dijk-Looyaard, Drinking-Water Research Scientist, National Institute of Public Health and Environmental Protection, Bilthoven, Netherlands

J.K. Fawell, Principal Toxicologist, Water Research Centre, Medmenham, England (*Rapporteur*)

J. Forslund, National Agency of Environmental Protection, Copenhagen, Denmark

E. Funari, Department of Environmental Hygiene, Istituto Superiore di Sanità, Rome, Italy

K. Khanna, Pharmacologist, Health Effects Branch, Office of Drinking-Water, United States Environmental Protection Agency, Washington, DC, USA

R. van Leeuwen, Toxicologist, National Institute of Public Health and Environmental Protection, Bilthoven, Netherlands

U. Lund, Head, Department of Chemistry, Water Quality Institute, Horsholm, Denmark

# ANNEX 1

M.E. Meek, Head, Priority Substances Section, Environmental Health Centre, Health and Welfare Canada, Ottawa, Canada

T. Ookubo, Head, Water Quality Examination Laboratory, Hachinohe Regional Water Supply Cooperation, Hachinohe, Japan

E. Sandberg, Toxicologist, National Food Administration, Uppsala, Sweden

U. Schlosser, Research Institute for Hygiene and Microbiology, Bad Elster, Germany

E.A. Simpson, Commission of the European Communities, Brussels, Belgium

J.A. Sokal, Head, Department of Toxicity Evaluation, Institute of Occupational Medicine, Lodz, Poland (*Chairman*)

M. Takeda, Director of Environmental Chemistry, National Institute of Hygienic Science, Tokyo, Japan

## *Observer*

A. Carlsen, Ministry of the Environment, National Agency of Environmental Protection, Miljöstyrelsen, Copenhagen, Denmark

## *Secretariat*

P. Bérubé, Programme Assistant, International Water Decade, Environment and Health, WHO Regional Office for Europe, Copenhagen, Denmark

O. Espinoza, Regional Officer for International Water Decade, Environment and Health, WHO Regional Office for Europe, Copenhagen, Denmark

D. Kello, Project Officer for Toxicology and Food Safety, Environment and Health, WHO Regional Office for Europe, Copenhagen, Denmark

D. Schutz, International Programme on Chemical Safety, Division of Environmental Health, World Health Organization, Geneva, Switzerland

S. Tarkowski, Director, Environment and Health, WHO Regional Office for Europe, Copenhagen, Denmark

J. Wilbourn, Unit of Carcinogen Identification and Evaluation, International Agency for Research on Cancer, Lyon, France

## First Review Group Meeting on Inorganics (Bilthoven, Netherlands, 18–22 March 1991)

### *Members*

E.A. Bababumni, Department of Biochemistry, University of Ibadan, Ibadan, Nigeria

K.L. Bailey, Health Effects Assessment Section, Criteria and Standards Division, Office of Drinking-Water, United States Environmental Protection Agency, Washington, DC, USA

G.F. Craun, Chief Epidemiologist, United States Environmental Protection Agency, Washington, DC, USA

A.M. van Dijk-Looyaard, Drinking-Water Research Scientist, National Institute of Public Health and Environmental Protection, Bilthoven, Netherlands

J.K. Fawell, Principal Toxicologist, Water Research Centre, Medmenham, England (*Rapporteur*)

R. van Leeuwen, Toxicologist, National Institute of Public Health and Environmental Protection, Bilthoven, Netherlands (*Chairman*)

M.E. Meek, Head, Priority Substances Section, Environmental Health Centre, Health and Welfare Canada, Ottawa, Canada

E. Poulsen, Chief Adviser in Toxicology, Institute of Toxicology, National Food Agency, Soborg, Denmark

Y.A. Rakhmanin, Head of Laboratory, Ministry of Health of the USSR Academy of Medical Sciences, A.N. Sysin Institute of General and Communal Hygiene, Moscow, USSR

V.R. Rao, Assistant Director and Head, Department of Toxicology, The Haffkine Institute, Parel, Bombay, India

F.G.R. Reyes, Professor of Food Toxicology, Department of Food Science, State University of Campinas, Brazil

F. Sartor, Institute of Hygiene and Epidemiology, Ministry of Public Health and the Family, Brussels, Belgium

J.A. Sokal, Head, Department of Toxicity Evaluation, Institute of Occupational Medicine, Lodz, Poland

M. Takeda, Director of Environmental Chemistry, National Institute of Hygienic Science, Tokyo, Japan

## *Observers*

J. Forslund, National Agency of Environmental Protection, Copenhagen, Denmark

I. Harimaya, Director of Water Quality Research, Kobe, Japan

M. Minowa, Director of Epidemiology, Institute of Public Health, Ministry of Health and Welfare, Tokyo, Japan

E.A. Simpson, Commission of the European Communities, Brussels, Belgium

J.F.M. Versteegh, National Institute of Public Health and Environmental Protection, Bilthoven, Netherlands

V. Vignier, Société Lyonnaise des Eaux Dumez, International Centre for Research on Water and the Environment (CIRSEE), Le Pecq, France

*Secretariat*

B. Chen, International Programme on Chemical Safety, Division of Environmental Health, World Health Organization, Geneva, Switzerland

H. Galal-Gorchev, International Programme on Chemical Safety, Division of Environmental Health, World Health Organization, Geneva, Switzerland

## Second Review Group Meeting on Organics (Copenhagen, Denmark, 8–12 April 1991)

*Members*

K. Bergman, Toxicologist, Medical Products Agency, Division of Pharmacology, Uppsala, Sweden

A. Carlsen, National Agency of Environmental Protection, Copenhagen, Denmark

H.H. Dieter, Director and Professor, Institute for Water, Soil and Air Hygiene of the Federal Office of Health, Berlin, Germany

P.M. Dudermel, Pasteur Institute, Lille, France

J.K. Fawell, Principal Toxicologist, Water Research Centre, Medmenham, England (*Rapporteur*)

J. Forslund, National Agency of Environmental Protection, Copenhagen, Denmark

R. Hasegawa, Section Chief, Division of Toxicology, National Institute of Hygienic Science, Tokyo, Japan

K. Hughes, Chemical Health Hazard Evaluator, Environmental Health Centre, Health and Welfare Canada, Ottawa, Canada

R. van Leeuwen, Toxicologist, National Institute of Public Health and Environmental Protection, Bilthoven, Netherlands

U. Lund, Head, Department of Chemistry, Water Quality Institute, Horsholm, Denmark

A. Patel, Toxicologist, Water Research Centre, Medmenham, England

Y. Richard, Chief, Department of Chemical Research, Société Degrémont, Rueil-Malmaison, France

E. Sandberg, Toxicologist, National Food Administration, Uppsala, Sweden

J.A. Sokal, Head, Department of Toxicity Evaluation, Institute of Occupational Medicine, Lodz, Poland (*Chairman*)

### Secretariat

X. Bonnefoy, Acting Regional Officer for Health Planning/Ecology, WHO Regional Office for Europe, Copenhagen, Denmark

H. Galal-Gorchev, International Programme on Chemical Safety, Division of Environmental Health, World Health Organization, Geneva, Switzerland

J. Gents, Secretary, International Water Decade, Environment and Health, WHO Regional Office for Europe, Copenhagen, Denmark

D. Kello, Project Officer for Toxicology and Food Safety, Environment and Health, WHO Regional Office for Europe, Copenhagen, Denmark

S. Tarkowski, Director, Environment and Health, WHO Regional Office for Europe, Copenhagen, Denmark

## Coordination Group Consultation (Geneva, Switzerland, 13–14 May 1991)

### Members

J. K. Fawell, Principal Toxicologist, Water Research Centre, Medmenham, England

J.R. Hickman, Director-General, Environmental Health Directorate, Health and Welfare Canada, Ottawa, Canada (*Moderator*)

U. Lund, Head, Department of Chemistry, Water Quality Institute, Horsholm, Denmark

B. Mintz, Chief, Health Effects Assessment Section, Criteria and Standards Division, Office of Drinking-Water, United States Environmental Protection Agency, Washington, DC, USA

E.B. Pike, Water Research Centre, Medmenham, England

### Secretariat

X. Bonnefoy, Acting Regional Officer for Health Planning/Ecology, WHO Regional Office for Europe, Copenhagen, Denmark (*Co-Rapporteur*)

H. Galal-Gorchev, International Programme on Chemical Safety, Division of Environmental Health, World Health Organization, Geneva, Switzerland (*Co-Rapporteur*)

R. Helmer, Prevention of Environmental Pollution, Division of Environmental Health, World Health Organization, Geneva, Switzerland

J. Kenny, Prevention of Environmental Pollution, Division of Environmental Health, World Health Organization, Geneva, Switzerland

M. Mercier, Manager, International Programme on Chemical Safety, Division of Environmental Health, World Health Organization, Geneva, Switzerland

G. Ozolins, Manager, Prevention of Environmental Pollution, Division of Environmental Health, World Health Organization, Geneva, Switzerland

P. Waight, Prevention of Environmental Pollution, Division of Environmental Health, World Health Organization, Geneva, Switzerland

## Review Group on Disinfectants and Disinfectant By-products (Bethesda, MD, USA, 10-14 June 1991)

### Members

H. Abouzaid, Chief, Water Quality Control Division, National Agency for Drinking-Water, Rabat-Chellah, Morocco

W. Almeida, Department of Preventive Medicine, State University of Campinas, Campinas, Brazil

M. Ando, National Institute of Hygienic Science, Division of Environmental Chemistry, Tokyo, Japan

R. Bull, Pharmacology/Toxicology Graduate Program, College of Pharmacy, Washington State University, Pullman, WA, USA

G. Burin, United States Environmental Protection Agency, Washington, DC, USA (*Vice-Chairman*)

J.K. Fawell, Principal Toxicologist, Water Research Centre, Medmenham, England (*Co-Rapporteur*)

B. Havlik, Institute of Hygiene and Epidemiology, Prague, Czechoslovakia

N. Mahabhol, Ministry of Public Health, Bangkok, Thailand

M.E. Meek, Head, Priority Substances Section, Environmental Health Centre, Health and Welfare Canada, Ottawa, Canada (*Co-Rapporteur*)

B. Mintz, Chief, Health Effects Assessment Section, Criteria and Standards Division, Office of Drinking-Water, United States Environmental Protection Agency, Washington, DC, USA (*Chairman*)

R. Packham, Marlow, England

J.F.M. Versteegh, National Institute of Public Health and Environmental Protection, Bilthoven, Netherlands

Z. Zholdakova, Academy of Medical Sciences, A.N. Sysin Institute of General and Communal Hygiene, Moscow, USSR

### Observers

J. Forslund, National Agency of Environmental Protection, Copenhagen, Denmark

E. Ohanian, Office of Science and Technology, United States Environmental Protection Agency, Washington, DC, USA

H. Sasaki, Water Quality Laboratory, Sapporro, Hokkaido, Japan

### Secretariat

R. Cantilli, United States Environmental Protection Agency, Washington, DC, USA

N. Chiu, United States Environmental Protection Agency, Washington, DC, USA

J. Du, United States Environmental Protection Agency, Washington, DC, USA

H. Galal-Gorchev, International Programme on Chemical Safety, Division of Environmental Health, World Health Organization, Geneva, Switzerland

J. Orme, Office of Science and Technology, United States Environmental Protection Agency, Washington, DC, USA

## Review Meeting on Pathogenic Agents and Volume 3 on Surveillance of Community Supplies (Harare, Zimbabwe, 24–28 June 1991)

### Members

H. Abouzaid, Chief, Water Quality Control Division, National Agency for Drinking-Water, Rabat-Chellah, Morocco

M.T. Boot, Programme Officer, IRC International Water and Sanitation Centre, The Hague, Netherlands

J.Z. Boutros, Consultant in Food and Water Control, Khartoum, Sudan (*Rapporteur*)

W. Fellows, Programme Officer, Water and Environmental Sanitation, UNICEF, Harare, Zimbabwe

F.J. Gumbo, Head of Water Laboratories, Operation, Maintenance and Water Laboratories Division, Ministry of Water (MAJI), Dar-es-Salaam, United Republic of Tanzania

A.H. Havelaar, National Institute of Public Health and Environmental Protection, Bilthoven, Netherlands

J. Hubley, Senior Lecturer in Health Education, Health Education Unit, Faculty of Health and Social Care, Leeds Polytechnic, Leeds, England

B. Jackson, Senior Engineering Advisor, British Development Division in East Africa, Nairobi, Kenya

E. Khaka, Ministry of Energy and Water Resources Development, Harare, Zimbabwe

S. Laver, Lecturer, Department of Community Medicine, University of Zimbabwe, Mount Pleasant, Harare, Zimbabwe

M.T. Martins, Associate Professor, Environmental Microbiology Laboratory, University of São Paulo, Brazil

P. Morgan, Advisor, Water and Sanitation, Ministry of Health, Blair Research Laboratory, Harare, Zimbabwe

S. Mtero, Principal Medical Research Officer, Ministry of Health, Blair Research Laboratory, Harare, Zimbabwe

S. Musingarabwi, Director, Environmental Health Services, Ministry of Health, Harare, Zimbabwe (*Vice-Chairman*)

F. Niang, Chief, Laboratory Service, Senegalese National Water Management Company, Dakar, Senegal

E.B. Pike, Water Research Centre, Medmenham, England

P.K. Ray, Director, Industrial Toxicology Research Centre, Lucknow, India

P. Taylor, Director, Training Centre for Water and Sanitation, Department of Civil Engineering, University of Zimbabwe, Harare, Zimbabwe (*Chairman*)

H. Utkilen, Scientist, National Institute of Public Health, Department of Environmental Medicine, Oslo, Norway

## *Observers*

M. Ellis, Primary Health Consultant, The Robens Institute of Health and Safety, University of Surrey, Guildford, England

D. Tolson, Aid Secretary, British High Commission, Harare, Zimbabwe

## *Secretariat*

J. Bartram, Manager, Overseas Development, The Robens Institute of Health and Safety, University of Surrey, Guildford, England

H. Galal-Gorchev, International Programme on Chemical Safety, Division of Environmental Health, World Health Organization, Geneva, Switzerland

R. Helmer, Prevention of Environmental Pollution, Division of Environmental Health, World Health Organization, Geneva, Switzerland

J. Kenny, Prevention of Environmental Pollution, Division of Environmental Health, World Health Organization, Geneva, Switzerland

V. Larby, The Robens Institute of Health and Safety, University of Surrey, Guildford, England

B. Lloyd, Head, Environmental Health, The Robens Institute of Health and Safety, University of Surrey, Guildford, England

K. Wedgwood, Research Officer, The Robens Institute of Health and Safety, University of Surrey, Guildford, England

F. Zawide, WHO Sanitary Engineer, Sub-region III, Harare, Zimbabwe

## Second Review Group Meeting on Pesticides (Rennes, France, 2–6 September 1991)

### Members

G. Burin, Toxicologist, United States Environmental Protection Agency, Washington, DC, USA (*Co-Rapporteur*)

A. Bruchet, Société Lyonnaise des Eaux Dumez, International Centre for Research on Water and the Environment (CIRSEE), Le Pecq, France

H.H. Dieter, Director and Professor, Institute for Water, Soil and Air Hygiene of the Federal Office of Health, Berlin, Germany

P.M. Dudermel, Pasteur Institute, Lille, France

J.K. Fawell, Principal Toxicologist, Water Research Centre, Medmenham, England (*Co-Rapporteur*)

J. Forslund, National Agency of Environmental Protection, Copenhagen, Denmark

E. Funari, Department of Environmental Hygiene, Istituto Superiore di Sanità, Rome, Italy

R. Halperin, Chief Engineer for Environmental Health, Ministry of Health, Jerusalem, Israel

K. Hughes, Chemical Health Hazard Evaluator, Priority Substances Section, Environmental Substances Division, Environmental Health Directorate, Environmental Health Centre, Ottawa, Canada

S. Kojima, Director of Environmental Chemistry, National Institute of Hygienic Science, Tokyo, Japan

A.M. Mahfouz, Senior Toxicologist and Pesticides Team Leader, Office of Science and Technology, United States Environmental Protection Agency, Washington, DC, USA

A. Montiel, Water Quality Control Officer, Water Management Company of Paris, Paris, France (*Chairman*)

E. Poulsen, Chief Adviser in Toxicology, Institute of Toxicology, National Food Agency, Soborg, Denmark

R. Seux, National School of Public Health, Rennes, France

E. Simpson, Commission of the European Communities, Brussels, Belgium

ANNEX 1

## Observers

M.J. Carroll, Area Registration Manager, Monsanto Services International, Brussels, Belgium

A. Hirata, Chief, Monitoring Section, Water Quality Management, Waterworks Bureau, Tokyo Metropolitan Government, Tokyo, Japan

H.P. Nigitz, Head, Regulatory Affairs, Agrolinz Agricultural Chemicals, Linz, Austria

E. Puri, Toxicologist, CIBA-GEIGY Ltd, Basel, Switzerland

G.A. Willis, Manager, Product Safety, ICI Agrochemicals, Fernhurst, Haslemere, Surrey, England

## Secretariat

X. Bonnefoy, Regional Adviser, Health Planning/Ecology, WHO Regional Office for Europe, Copenhagen, Denmark

H. Galal-Gorchev, International Programme on Chemical Safety, Division of Environmental Health, World Health Organization, Geneva, Switzerland

J. Gents, Programme Secretary, Environment and Health, WHO Regional Office for Europe, Copenhagen, Denmark

## Second Review Group Meeting on Inorganics (Brussels, Belgium, 14–18 October 1991)

## Members

Y. Aida, Senior Research Scientist, Division of Risk Assessment, National Institute of Hygienic Science, Kamiyoga, Setagayaku, Tokyo, Japan

J. Alexander, Deputy Director, Department of Environmental Medicine, National Institute of Public Health, Oslo, Norway

K.L. Bailey, Health Effects Assessment Section, Criteria and Standards Division, Office of Drinking-Water, United States Environmental Protection Agency, Washington, DC, USA

H.H. Dieter, Director and Professor, Toxicologist, Institute for Water, Soil and Air Hygiene of the Federal Office of Health, Berlin, Germany

J.K. Fawell, Principal Toxicologist, Water Research Centre, Medmenham, England (*Co-Rapporteur*)

A. Lafontaine, Honorary Director, Institute of Hygiene and Epidemiology, Brussels, Belgium

M.E. Meek, Head, Priority Substances Section, Environmental Health Centre, Health and Welfare Canada, Ottawa, Canada

B. Naima, Director, Water Quality Laboratory, National Agency for Drinking-Water, Rabat-Chellah, Morocco

G.D. Nielsen, Department of Environmental Medicine, Odense University, Odense, Denmark

R.F. Packham, Marlow, England

Y.A. Rakhmanin, Head of Laboratory, Ministry of Health of the USSR Academy of Medical Sciences, A.N. Sysin Institute of General and Communal Hygiene, Moscow, USSR

Tharwat Saleh, Project Manager, WHO Project EFY/CWS/002, Cairo, Egypt

R. Sarin, Assistant Director, Scientist and Head, Basic Research Division, National Environmental Engineering Research Institute (NEERI), Nehru Marg, Nagpur, India

F. Sartor, Institute of Hygiene and Epidemiology, Ministry of Public Health and the Family, Brussels, Belgium (*Chairman*)

J.F.M. Versteegh, National Institute of Public Health and Environmental Protection, Bilthoven, Netherlands

### Observers

J. Forslund, National Agency of Environmental Protection, Copenhagen, Denmark

E.A. Simpson, Commission of the European Communities, Brussels, Belgium

V. Vignier, Société Lyonnaise des Eaux Dumez, International Centre for Research on Water and the Environment (CIRSEE), Le Pecq, France

### Secretariat

X. Bonnefoy, Regional Adviser, Health Planning/Ecology, WHO Regional Office for Europe, Copenhagen, Denmark

H. Galal-Gorchev, International Programme on Chemical Safety, Division of Environmental Health, World Health Organization, Geneva, Switzerland (*Co-Rapporteur*)

C. Martin, Prevention of Environmental Pollution, Division of Environmental Health, World Health Organization, Geneva, Switzerland

## Radionuclides Meeting (Medmenham, England, 22–24 January 1992)

### Members

O. Hydes, Drinking-Water Inspectorate, Department of the Environment, London, England

D.P. Meyerhof, Bureau of Radiation and Medical Devices, Department of National Health and Welfare, Ottawa, Canada

ANNEX 1

J.-C. Nénot, Director of Research, Institute for Nuclear Protection and Safety, Fontenay-aux-Roses, France

K.C. Pillai, Health Physics Division, Bhabha Atomic Research Centre, Bombay, India

A. Randell, Senior Officer, Food Quality and Standards Service, Food and Agriculture Organization of the United Nations, Rome, Italy

C. Robinson, National Radiological Protection Board, Chilton, Didcot, England (*Co-Rapporteur*)

L.B. Sztanyik, Director, "Frédéric Joliot-Curie" National Research Institute for Radiobiology and Radiohygiene, Budapest, Hungary (*Chairman*)

E. Wirth, Institute for Radiation Hygiene, Federal Office for Radiation Protection, Neuerberg, Germany

### *Secretariat*

P.J. Waight, Prevention of Environmental Pollution, Division of Environmental Health, World Health Organization, Geneva, Switzerland (*Co-Rapporteur*)

## Analytical and Treatment Methods (Medmenham, England, 27–29 January 1992)

### *Members*

H. Abouzaid, Chief, Water Quality Control Division, National Agency for Drinking-Water, Rabat-Chellah, Morocco

S. Clark, Chief, Drinking Water Technology Branch, Office of Groundwater and Drinking Water, United States Environmental Protection Agency, Washington, DC, USA

A.M. van Dijk-Looyaard, Drinking-Water Research Scientist, National Institute of Public Health and Environmental Protection, Bilthoven, Netherlands

J. Forslund, National Agency of Environmental Protection, Copenhagen, Denmark

D. Green, Criteria Section, Environmental Health Centre, Department of National Health and Welfare, Ottawa, Canada (*Co-Rapporteur*)

I. Licsko, Research Centre for Water Resources Development (VITUKI), Budapest, Hungary

B. Lloyd, Head, Environmental Health, The Robens Institute of Health and Safety, University of Surrey, Guildford, England

D.P. Meyerhof, Bureau of Radiation and Medical Devices, Department of National Health and Welfare, Ottawa, Canada

A. Montiel, Water Quality Control Officer, Water Management Company of Paris, Paris, France (*Co-Rapporteur*)

R.F. Packham, Marlow, England (*Chairman*)

R. Sarin, Assistant Director, Scientist and Head, Basic Research Division, National Environmental Engineering Research Institute (NEERI), Nehru Marg, Nagpur, India

## Observers

T. Aizawa, Department of Sanitary Engineering, Institute of Public Health, Tokyo, Japan

R.A. Breach, Water Quality Manager, Severn Trent Water, Birmingham, England

O. Hydes, Drinking Water Inspectorate, Department of the Environment, London, England

M. Ichinohe, Bureau of Waterworks, Tokyo Metropolitan Government, Tokyo, Japan

E. Simpson, Commission of the European Communities, Brussels, Belgium

M. Tsuji, Ministry of Health and Welfare, Tokyo, Japan

## Secretariat

X. Bonnefoy, Regional Adviser, Health Planning/Ecology, WHO Regional Office for Europe, Copenhagen, Denmark

B. Crathorne, Water Research Centre, Medmenham, England

J.K. Fawell, Principal Toxicologist, Water Research Centre, Medmenham, England

H. Galal-Gorchev, International Programme on Chemical Safety, Division of Environmental Health, World Health Organization, Geneva, Switzerland

E.B. Pike, Water Research Centre, Medmenham, England

# WHO Consolidation Meeting on Organics and Pesticides (Medmenham, England, 30–31 January 1992)

## Members

J.K. Fawell, Principal Toxicologist, Water Research Centre, Medmenham, England

R. van Leeuwen, Toxicologist, National Institute of Public Health and Environmental Protection, Bilthoven, Netherlands (*Moderator*)

U. Lund, Head, Department of Chemistry, Water Quality Institute, Horsholm, Denmark

M. Sheffer, Scientific Editor, Orleans, Canada

## Secretariat

X. Bonnefoy, Regional Adviser, Health Planning/Ecology, WHO Regional Office for Europe, Copenhagen, Denmark (*Co-Rapporteur*)

H. Galal-Gorchev, International Programme on Chemical Safety, Division of Environmental Health, World Health Organization, Geneva, Switzerland (*Co-Rapporteur*)

## Final Drafts Preparation Meeting for Volumes 1 and 2 (Val David, Quebec, Canada, 19-22 May 1992)

### Members

K. Bentley, Director, Environmental Health, Health Advancement Division, Australian Department of Health, Housing and Community Services, Woden, Australia

J.K. Fawell, Principal Toxicologist, Water Research Centre, Medmenham, England

J.R. Hickman, Director-General, Environmental Health Directorate, Department of National Health and Welfare, Ottawa, Canada (*Chairman*)

U. Lund, Head, Department of Chemistry, Water Quality Institute, Horsholm, Denmark

M.E. Meek, Head, Priority Substances Section, Environmental Health Centre, Department of National Health and Welfare, Ottawa, Canada

B. Mintz, Chief, Health Effects Assessment Section, Criteria and Standards Division, Office of Drinking-Water, United States Environmental Protection Agency, Washington, DC, USA

R.F. Packham, Marlow, England

E.B. Pike, Water Research Centre, Medmenham, England

M. Sheffer, Scientific Editor, Orleans, Canada

P. Toft, Health Protection Branch, Environmental Health Directorate, Department of National Health and Welfare, Ottawa, Canada

G. Wood, Health Protection Branch, Environmental Health Directorate, Department of National Health and Welfare, Ottawa, Canada

### Secretariat

X. Bonnefoy, Regional Adviser, Health Planning/Ecology, WHO Regional Office for Europe, Copenhagen, Denmark (*Co-Rapporteur*)

H. Galal-Gorchev, International Programme on Chemical Safety, World Health Organization, Geneva, Switzerland (*Co-Rapporteur*)

G. Ozolins, Manager, Prevention of Environmental Pollution, Division of Environmental Health, World Health Organization, Geneva, Switzerland

## Final Task Group Meeting (Geneva, Switzerland, 21–25 September 1992)

### Members*

H. Abouzaid, Chief, Water Quality Control Division, National Agency for Drinking-Water, Rabat-Chellah, Morocco

M. Aguilar, Director of Basic Sanitation, Department of Environmental and Occupational Health and Basic Sanitation, Mexico City, Mexico

J. Alexander, Deputy Director, Department of Environmental Medicine, National Institute of Public Health, Oslo, Norway

V. Angjeli, Chief of Communal Hygiene Division, Research Institute of Hygiene and Epidemiology, Tirana, Albania

L. Anukam, Federal Environmental Protection Agency (FEPA), Department of Planning and Evaluation, Federal Secretariat Complex (Phase II), Ikoyi, Lagos, Nigeria

W.S. Assoy, Director, Environmental Health Service, Department of Health, Manila, Philippines

Changjie Chen, Director, Institute of Environmental Health Monitoring, Chinese Academy of Preventive Medicine, Beijing, China

M. Csanady, Department Leader, National Institute of Hygiene, Budapest, Hungary

H.H. Dieter, Director and Professor, Institute for Water, Soil and Air Hygiene of the Federal Office of Health, Berlin, Germany

F.K. El Jack, Head of Water Department, National Chemical Laboratories, Khartoum, Sudan

J. Forslund, National Agency of Environmental Protection, Copenhagen, Denmark (*Vice-Chairman*)

E. Funari, Department of Environmental Hygiene, Istituto Superiore di Sanità, Rome, Italy

E. Gonzalez, Chief, Department of Water Quality, Water Supply and Sewerage, San José, Costa Rica

F.J. Gumbo, Head of Water Laboratories, Operation, Maintenance and Water Laboratories Division, Ministry of Water (MAJI), Dar-es-Salaam, United Republic of Tanzania

B. Havlik, Head of Water Hygiene Branch, National Institute of Public Health, Prague, Czechoslovakia

H.M.S.S.D. Herath, Deputy Director General, Public Health Services, Ministry of Health, Colombo, Sri Lanka

---

* Invited but unable to attend: Director-General of Health, Islamabad, Pakistan; F. Sartor, Institute of Hygiene and Epidemiology, Ministry of Public Health and the Family, Brussels, Belgium.

# ANNEX 1

L. Hiisvirta, Chief Engineer, Ministry of Social Affairs and Health, Helsinki, Finland

J. Kariuki, Senior Public Health Officer, Division of Environmental Health, Ministry of Health, Nairobi, Kenya

M. Kitenge, Director, Department of Local Production Control, Zaire Control Agency, Kinshasa, Zaire

F.X.R. van Leeuwen, Senior Toxicologist, National Institute of Public Health and Environmental Protection, Bilthoven, Netherlands

Y. Magara, Director, Department of Water Supply Engineering, Institute of Public Health, Tokyo, Japan

N.S. McDonald, Director, Water Branch, Department of Primary Industries and Energy, Canberra, Australia

B. Mintz, Chief, Exposure Assessment and Environmental Fate Section, Office of Science and Technology, United States Environmental Protection Agency, Washington, DC, USA

F. Niang, Chief, Laboratory Service, Senegalese National Water Management Company, Dakar, Senegal

R.F. Packham, Marlow, England

Y.A. Rakhmanin, Academician of Russian Academy of Natural Sciences, A.N. Sysin Research Institute of Human Ecology and Environmental Health, Moscow, Russian Federation

F.G.R. Reyes, Professor of Food Toxicology, Department of Food Science, State University of Campinas, Brazil (*Rapporteur*)

T. Saleh, WHO Regional Support Office, Cairo, Egypt

E. Sandberg, Senior Toxicologist, National Food Administration, Uppsala, Sweden

Nantana Santatiwut, Director, Environmental Health Division, Department of Health, Ministry of Public Health, Bangkok, Thailand

R. Sarin, Scientist, National Environmental Engineering Research Institute (NEERI), Nehru Marg, Nagpur, India

C. Shaw, Senior Advisor Scientist, Public Health Services, Department of Health, Wellington, New Zealand

J.A. Sokal, Director, Institute of Occupational Medicine and Environmental Health, Sosnowiec, Poland

P. Toft, Health Protection Branch, Environmental Health Directorate, Department of National Health and Welfare, Ottawa, Canada (*Chairman*)

D. Tricard, Sanitary Engineer, Ministry of Health and Humanitarian Action, Department of Health, Paris, France

## Observers

M.J. Crick, Radiation Safety Specialist, International Atomic Energy Agency, Vienna, Austria

A.M. van Dijk-Looyaard, Senior Scientist Drinking-Water Standards, KIWA N.V. Research and Consultancy, Nieuwegein, Netherlands

O. Hydes, Drinking Water Inspectorate, Department of the Environment, London, England

M. Rapinat, International Water Supply Association, Compagnie générale des Eaux, Paris, France

Y. Richard, Head Engineer, Société DEGREMONT-CIRSEE, Le Pecq, France

H. Rousseau, Division des Eaux de Consommation, Direction des Ecosystèmes urbains, Ministère de l'Environnement, Ste Foy, Québec, Canada

J.E. Samdal, Norwegian Institute for Water Research (NIVA), Oslo, Norway

F. Sarhan, CIBA-GEIGY Ltd, Basel, Switzerland (representing the International Group of National Associations of Manufacturers of Agrochemical Products)

E.A. Simpson, Commission of the European Communities, Brussels, Belgium

T. Yanagisawa, Director, Technical Management Section, Management and Planning Division, Bureau of Waterworks, Tokyo, Japan

## Secretariat

J. Bartram, Manager, Overseas Development, The Robens Institute of Health and Safety, University of Surrey, Guildford, England

X. Bonnefoy, Environmental Health Planning/Ecology, WHO Regional Office for Europe, Copenhagen, Denmark

A. Enevoldsen, Environmental Health Planning/Ecology, WHO Regional Office for Europe, Copenhagen, Denmark

J.K. Fawell, Principal Toxicologist, Water Research Centre, Medmenham, England

B.H. Fenger, Water and Waste Scientist, WHO European Office for Environment and Health, Rome, Italy

H. Galal-Gorchev, International Programme on Chemical Safety, World Health Organization, Geneva, Switzerland

R. Helmer, Prevention of Environmental Pollution, Division of Environmental Health, World Health Organization, Geneva, Switzerland

J. Kenny, Prevention of Environmental Pollution, Division of Environmental Health, World Health Organization, Geneva, Switzerland

U. Lund, Water Quality Institute, Horsholm, Denmark

# ANNEX 1

M. Mercier, Director, International Programme on Chemical Safety, World Health Organization, Geneva, Switzerland

H. Moller, Scientist, Unit of Carcinogen Identification and Evaluation, International Agency for Research on Cancer, Lyon, France

G. Ozolins, Manager, Prevention of Environmental Pollution, Division of Environmental Health, World Health Organization, Geneva, Switzerland

E.B. Pike, Water Research Centre, Medmenham, England

M. Sheffer, Scientific Editor, Orleans, Canada

S. Tarkowski, Director, Environment and Health, WHO Regional Office for Europe, Copenhagen, Denmark

Annex 2
# Tables of guideline values

The following tables present a summary of guideline values for microorganisms and chemicals in drinking-water. Individual values should not be used directly from the tables. The guideline values must be used and interpreted in conjunction with the information contained in the text and in Volume 2, *Health criteria and other supporting information*.

## Table A2.1. Bacteriological quality of drinking-water[a]

| Organisms | Guideline value |
|---|---|
| **All water intended for drinking** | |
| E. coli or thermotolerant coliform bacteria[b,c] | Must not be detectable in any 100-ml sample |
| **Treated water entering the distribution system** | |
| E. coli or thermotolerant coliform bacteria[b] | Must not be detectable in any 100-ml sample |
| Total coliform bacteria | Must not be detectable in any 100-ml sample |
| **Treated water in the distribution system** | |
| E. coli or thermotolerant coliform bacteria[b] | Must not be detectable in any 100-ml sample |
| Total coliform bacteria | Must not be detectable in any 100-ml sample. In the case of large supplies, where sufficient samples are examined, must not be present in 95% of samples taken throughout any 12-month period |

[a] Immediate investigative action must be taken if either E. coli or total coliform bacteria are detected. The minimum action in the case of total coliform bacteria is repeat sampling; if these bacteria are detected in the repeat sample, the cause must be determined by immediate further investigation.

[b] Although E. coli is the more precise indicator of faecal pollution, the count of thermotolerant coliform bacteria is an acceptable alternative. If necessary, proper confirmatory tests must be carried out. Total coliform bacteria are not acceptable indicators of the sanitary quality of rural water supplies, particularly in tropical areas where many bacteria of no sanitary significance occur in almost all untreated supplies.

[c] It is recognized that, in the great majority of rural water supplies in developing countries, faecal contamination is widespread. Under these conditions, the national surveillance agency should set medium-term targets for the progressive improvement of water supplies, as recommended in Volume 3 of *Guidelines for drinking-water quality*.

## Table A2.2. Chemicals of health significance in drinking-water

### A. Inorganic constituents

|  | Guideline value (mg/litre) | Remarks |
|---|---|---|
| antimony | 0.005 (P)[a] | |
| arsenic | 0.01[b](P) | For excess skin cancer risk of $6 \times 10^{-4}$ |
| barium | 0.7 | |
| beryllium | | NAD[c] |
| boron | 0.3 | |
| cadmium | 0.003 | |
| chromium | 0.05 (P) | |
| copper | 2 (P) | ATO[d] |
| cyanide | 0.07 | |
| fluoride | 1.5 | Climatic conditions, volume of water consumed, and intake from other sources should be considered when setting national standards |
| lead | 0.01 | It is recognized that not all water will meet the guideline value immediately; meanwhile, all other recommended measures to reduce the total exposure to lead should be implemented |
| manganese | 0.5 (P) | ATO |
| mercury (total) | 0.001 | |
| molybdenum | 0.07 | |
| nickel | 0.02 | |
| nitrate (as $NO_3^-$) | 50 | The sum of the ratio of the concentration of each to its respective guideline value should not exceed 1 |
| nitrite (as $NO_2^-$) | 3 (P) | |
| selenium | 0.01 | |
| uranium | | NAD |

## B. Organic constituents

| | Guideline value (μg/litre) | Remarks |
|---|---|---|
| *Chlorinated alkanes* | | |
| carbon tetrachloride | 2 | |
| dichloromethane | 20 | |
| 1,1-dichloroethane | | NAD |
| 1,2-dichloroethane | 30[b] | for excess risk of $10^{-5}$ |
| 1,1,1-trichloroethane | 2000 (P) | |
| *Chlorinated ethenes* | | |
| vinyl chloride | 5[b] | for excess risk of $10^{-5}$ |
| 1,1-dichloroethene | 30 | |
| 1,2-dichloroethene | 50 | |
| trichloroethene | 70 (P) | |
| tetrachloroethene | 40 | |
| *Aromatic hydrocarbons* | | |
| benzene | 10[b] | for excess risk of $10^{-5}$ |
| toluene | 700 | ATO |
| xylenes | 500 | ATO |
| ethylbenzene | 300 | ATO |
| styrene | 20 | ATO |
| benzo[a]pyrene | 0.7[b] | for excess risk of $10^{-5}$ |
| *Chlorinated benzenes* | | |
| monochlorobenzene | 300 | ATO |
| 1,2-dichlorobenzene | 1000 | ATO |
| 1,3-dichlorobenzene | | NAD |
| 1,4-dichlorobenzene | 300 | ATO |
| trichlorobenzenes (total) | 20 | ATO |
| *Miscellaneous* | | |
| di(2-ethylhexyl)adipate | 80 | |
| di(2-ethylhexyl)phthalate | 8 | |
| acrylamide | 0.5[b] | for excess risk of $10^{-5}$ |
| epichlorohydrin | 0.4 (P) | |
| hexachlorobutadiene | 0.6 | |
| edetic acid (EDTA) | 200 (P) | |
| nitrilotriacetic acid | 200 | |
| dialkyltins | | NAD |
| tributyltin oxide | 2 | |

GUIDELINES FOR DRINKING-WATER QUALITY

## C. Pesticides

| | Guideline value (µg/litre) | Remarks |
|---|---|---|
| alachlor | 20[b] | for excess risk of $10^{-5}$ |
| aldicarb | 10 | |
| aldrin/dieldrin | 0.03 | |
| atrazine | 2 | |
| bentazone | 30 | |
| carbofuran | 5 | |
| chlordane | 0.2 | |
| chlorotoluron | 30 | |
| DDT | 2 | |
| 1,2-dibromo 3-chloropropane | 1[b] | for excess risk of $10^{-5}$ |
| 2,4-D | 30 | |
| 1,2-dichloropropane | 20 (P) | |
| 1,3-dichloropropane | | NAD |
| 1,3-dichloropropene | 20[b] | for excess risk of $10^{-5}$ |
| ethylene dibromide | | NAD |
| heptachlor and heptachlor epoxide | 0.03 | |
| hexachlorobenzene | 1[b] | for excess risk of $10^{-5}$ |
| isoproturon | 9 | |
| lindane | 2 | |
| MCPA | 2 | |
| methoxychlor | 20 | |
| metolachlor | 10 | |
| molinate | 6 | |
| pendimethalin | 20 | |
| pentachlorophenol | 9 (P) | |
| permethrin | 20 | |
| propanil | 20 | |
| pyridate | 100 | |
| simazine | 2 | |
| trifluralin | 20 | |
| chlorophenoxy herbicides other than 2,4-D and MCPA | | |
| 2,4-DB | 90 | |
| dichlorprop | 100 | |
| fenoprop | 9 | |
| MCPB | | NAD |
| mecoprop | 10 | |
| 2,4,5-T | 9 | |

## D. Disinfectants and disinfectant by-products

| Disinfectants | Guideline value (mg/litre) | Remarks |
|---|---|---|
| monochloramine | 3 | |
| di- and trichloramine | | NAD |
| chlorine | 5 | ATO. For effective disinfection there should be a residual concentration of free chlorine of ⩾0.5 mg/litre after at least 30 minutes contact time at pH <8.0 |
| chlorine dioxide | | A guideline value has not been established because of the rapid breakdown of chlorine dioxide and because the chlorite guideline value is adequately protective for potential toxicity from chlorine dioxide |
| iodine | | NAD |

| Disinfectant by-products | Guideline value (µg/litre) | Remarks |
|---|---|---|
| bromate | 25[b] (P) | for $7 \times 10^{-5}$ excess risk |
| chlorate | | NAD |
| chlorite | 200 (P) | |
| chlorophenols | | |
|   2-chlorophenol | | NAD |
|   2,4-dichlorophenol | | NAD |
|   2,4,6-trichlorophenol | 200[b] | for excess risk of $10^{-5}$, ATO |
| formaldehyde | 900 | |
| MX | | NAD |
| trihalomethanes | | The sum of the ratio of the concentration of each to its respective guideline value should not exceed 1 |
|   bromoform | 100 | |
|   dibromochloromethane | 100 | |
|   bromodichloromethane | 60[b] | for excess risk of $10^{-5}$ |
|   chloroform | 200[b] | for excess risk of $10^{-5}$ |
| chlorinated acetic acids | | |
|   monochloroacetic acid | | NAD |
|   dichloroacetic acid | 50 (P) | |
|   trichloroacetic acid | 100 (P) | |
| chloral hydrate (trichloroacetaldehyde) | 10 (P) | |
| chloroacetone | | NAD |

| Disinfectant by-products | Guideline value (µg/litre) | Remarks |
|---|---|---|
| halogenated acetonitriles | | |
|   dichloroacetonitrile | 90 (P) | |
|   dibromoacetonitrile | 100 (P) | |
|   bromochloroacetonitrile | | NAD |
|   trichloroacetonitrile | 1 (P) | |
| cyanogen chloride (as CN) | 70 | |
| chloropicrin | | NAD |

[a] (P) — Provisional guideline value. This term is used for constituents for which there is some evidence of a potential hazard but where the available information on health effects is limited; or where an uncertainty factor greater than 1000 has been used in the derivation of the tolerable daily intake (TDI). Provisional guideline values are also recommended: (1) for substances for which the calculated guideline value would be below the practical quantification level, or below the level that can be achieved through practical treatment methods; or (2) where disinfection is likely to result in the guideline value being exceeded.

[b] For substances that are considered to be carcinogenic, the guideline value is the concentration in drinking-water associated with an excess lifetime cancer risk of $10^{-5}$ (one additional cancer per 100 000 of the population ingesting drinking-water containing the substance at the guideline value for 70 years). Concentrations associated with estimated excess lifetime cancer risks of $10^{-4}$ and $10^{-6}$ can be calculated by multiplying and dividing, respectively, the guideline value by 10.

    In cases in which the concentration associated with an excess lifetime cancer risk of $10^{-5}$ is not feasible as a result of inadequate analytical or treatment technology, a provisional guideline value is recommended at a practicable level and the estimated associated excess lifetime cancer risk presented.

    It should be emphasized that the guideline values for carcinogenic substances have been computed from hypothetical mathematical models that cannot be verified experimentally and that the values should be interpreted differently than TDI-based values because of the lack of precision of the models. At best, these values must be regarded as rough estimates of cancer risk. However, the models used are conservative and probably err on the side of caution. Moderate short-term exposure to levels exceeding the guideline value for carcinogens does not significantly affect the risk.

[c] NAD— No adequate data to permit recommendation of a health-based guideline value.

[d] ATO— Concentrations of the substance at or below the health-based guideline value may affect the appearance, taste, or odour of the water.

*Table A2.3. Chemicals not of health significance at concentrations normally found in drinking-water*

| Chemical | Remarks |
| --- | --- |
| asbestos | U |
| silver | U |
| tin | U |

U — It is unnecessary to recommend a health-based guideline value for these compounds because they are not hazardous to human health at concentrations normally found in drinking-water.

*Table A2.4. Radioactive constituents of drinking-water*

| | Screening value (Bq/litre) | Remarks |
| --- | --- | --- |
| gross alpha activity | 0.1 | If a screening value is exceeded, more detailed radionuclide analysis is necessary. Higher values do not necessarily imply that the water is unsuitable for human consumption |
| gross beta activity | 1 | |

**Table A2.5. Substances and parameters in drinking-water that may give rise to complaints from consumers**

| | Levels likely to give rise to consumer complaints[a] | Reasons for consumer complaints |
|---|---|---|
| *Physical parameters* | | |
| colour | 15 TCU[b] | appearance |
| taste and odour | — | should be acceptable |
| temperature | — | should be acceptable |
| turbidity | 5 NTU[c] | appearance; for effective terminal disinfection, median turbidity ≤1NTU, single sample ≤5NTU |
| *Inorganic constituents* | | |
| aluminium | 0.2 mg/l | depositions, discoloration |
| ammonia | 1.5 mg/l | odour and taste |
| chloride | 250 mg/l | taste, corrosion |
| copper | 1 mg/l | staining of laundry and sanitary ware (health-based provisional guideline value 2 mg/litre) |
| hardness | — | high hardness: scale deposition, scum formation |
| | | low hardness: possible corrosion |
| hydrogen sulfide | 0.05 mg/l | odour and taste |
| iron | 0.3 mg/l | staining of laundry and sanitary ware |
| manganese | 0.1 mg/l | staining of laundry and sanitary ware (health-based provisional guideline value 0.5 mg/litre) |
| dissolved oxygen | — | indirect effects |
| pH | — | low pH: corrosion |
| | | high pH: taste, soapy feel |
| | | preferably <8.0 for effective disinfection with chlorine |
| sodium | 200 mg/l | taste |
| sulfate | 250 mg/l | taste, corrosion |
| total dissolved solids | 1000 mg/l | taste |
| zinc | 3 mg/l | appearance, taste |
| *Organic constituents* | | |
| toluene | 24–170 µg/l | odour, taste (health-based guideline value 700 µg/l) |
| xylene | 20–1800 µg/l | odour, taste (health-based guideline value 500 µg/l) |
| ethylbenzene | 2–200 µg/l | odour, taste (health-based guideline value 300 µg/l) |
| styrene | 4–2600 µg/l | odour, taste (health-based guideline value 20 µg/l) |

# ANNEX 2

|  | Levels likely to give rise to consumer complaints[a] | Reasons for consumer complaints |
|---|---|---|
| monochlorobenzene | 10–120 µg/l | odour, taste (health-based guideline value 300 µg/l) |
| 1,2-dichlorobenzene | 1–10 µg/l | odour, taste (health-based guideline value 1000 µg/l) |
| 1,4-dichlorobenzene | 0.3–30 µg/l | odour, taste (health-based guideline value 300 µg/l) |
| trichlorobenzenes (total) | 5–50 µg/l | odour, taste (health-based guideline value 20 µg/l) |
| synthetic detergents | — | foaming, taste, odour |
| *Disinfectants and disinfectant by-products* | | |
| chlorine | 600–1000 µg/l | taste and odour (health-based guideline value 5 mg/l) |
| chlorophenols | | |
| 2-chlorophenol | 0.1–10 µg/l | taste, odour |
| 2,4-dichlorophenol | 0.3–40 µg/l | taste, odour |
| 2,4,6-trichlorophenol | 2–300 µg/l | taste, odour (health-based guideline value 200 µg/l) |

[a] The levels indicated are not precise numbers. Problems may occur at lower or higher values according to local circumstances. A range of taste and odour threshold concentrations is given for organic constituents.

[b] TCU, time colour unit.

[c] NTU, nephelometric turbidity unit.

# Index

*Acanthamoeba* 9, 13
Acceptability 4, 122-130
   disinfectants and disinfectant by-products 129-130, 181
   guideline values 180-181
   inorganic constituents 124-128, 180
   organic constituents 128-129, 180-181
   physical parameters 123-124, 180
Acceptable daily intake (ADI) 32
Acetic acids, chlorinated 101-103, 177
*Acinetobacter* 9
Acrylamide 72, 175
Actinomycetes 12
Adenoviruses 10
*Aeromonas* 9, 10, 12, 13, 127
Aggressivity index 141
Alachlor 76, 176
Aldicarb 76-77, 176
Aldrin 77, 176
Algae
   blooms 12
   toxins 9-11
Alkanes, chlorinated 57-60, 175
Alpha activity 116-117, 119, 120
   screening values 117, 179
Alpha-emitting radionuclides 118, 119
Alumina, activated 136
Aluminium 39-40, 124, 180
Alzheimer disease 39-40
Ammonia 40, 124, 180
*Anabaena* 11
Analytical methods
   chemicals 111-113
      accuracy requirements 111-112
      quality control 111, 113
      selection of suitable 112-113
   radionuclides 118-119
*Ancylostoma* 9
Animals
   nuisance 12
   reservoirs of infections 10-11
Anodic protection 139
Antimony 40-41, 174

*Aphanizomenon* 11
Argyria 55
Aromatic hydrocarbons 64-68, 175
Arsenic 41-42, 174
Asbestos 42, 179
Asbestos-cement (A/C) pipes 140
*Ascaris* 9
*Asellus aquaticus* 12
Atrazine 77-78, 176

Bacteria, pathogenic 8, 9, 10
Bacteriological quality guidelines 22
Bacteriophages 18
*Bacteroides fragilis* 18
*Balantidium coli* (balantidiasis) 9
Barium 42-43, 174
Bentazone 78, 176
Benzene 64-65, 175
Benzenes, chlorinated 68-70, 175
Benzo[*a*]pyrene 67-68, 175
Beryllium 43, 174
Beta activity 116-118, 119, 120
   screening values 117, 179
Beta-emitting radionuclides 117-118
Bifidobacteria 18
Bilharzia 9
Blue-green algae *see* Cyanobacteria
Body weight 30-31
Boiling of water 143
Bone-meal, charred 136
Boron 43-44, 174
Bromate 96, 177
Bromochloroacetonitrile 104, 178
Bromodichloromethane (BDCM) 99, 100-101, 177
Bromoform 99-100, 177
*Buttiauxella agrestis* 17

Cadmium 44, 174
Caesium-134 ($^{134}$Cs) 117, 118
Caesium-137 ($^{137}$Cs) 117, 118
Calcium carbonate scale 125, 141-142
Calcium sulfate 127
*Campylobacter coli* 8, 10

*Campylobacter jejuni* 8, 10
Cancer risk (*see also* Carcinogens)
 radiation-associated 115
Carbofuran 78-79, 176
Carbon-14 ($^{14}$C) 117
Carbon filters 135-136
Carbon tetrachloride 7, 57-58, 175
Carcinogens
 derivation of guideline values 35-38
 guideline values 178
 IARC classification 35, 36-37
Cardiovascular disease 42, 48
Cement pipes, corrosion 140-141
Chemicals 1-2, 30-113
 criteria for selection 6-7
 dermal absorption 31
 emergencies involving 143
 guideline values
  derivation methods 32-38
  tables 174-179
 health risk assessment 31-38
 health risks 3-4
 information sources 30
 inhalation 31
 inorganic 39-57, 124-128, 174, 180
 mixtures 39
 monitoring 105-113
 organic 57-75, 128-129, 175, 180-181
 summary statements 39-105
 water consumption data 30-31
*Chironomus* larvae 12
Chloral hydrate 103, 177
Chloramines 94, 135, 177
Chlorate 96, 177
Chlordane 79, 176
Chloride 45, 124-125, 180
 corrosion and 45, 139
Chlorine 7, 94-95
 acceptable levels 129, 181
 application 134, 135
 emergency disinfection 143
 guideline values 95, 177
 resistance of pathogens 10-11
Chlorine dioxide 95, 135, 177
Chlorite 96-97, 177
3-Chloro-4-dichloromethyl-5-hydroxy-2(5H)-
 furanone (MX) 98, 177
Chloroacetones 103, 177
Chloroform 1, 99, 101, 177
2-Chlorophenol 97, 130, 177, 181
Chlorophenols 97-98, 130, 177, 181
Chlorophenoxy herbicides 91-93, 176
Chloropicrin 105, 178
Chlorotoluron 79-80, 176
Chromium 45-46, 174

*Chydorus sphaericus* 12
*Citrobacter* 16
*Citrobacter freundii* 16-17
Clostridia, sulfite-reducing 15, 18
*Clostridium perfringens (welchii)* 18
Coagulation 134
Coal-tar pipe linings 68
Cobalt-60 ($^{60}$Co) 117, 118
Codex standards 6
Coliform bacteria 15-17, 19
 guideline values 22
 thermotolerant 15, 16, 20, 21
  guideline values 22, 173
 total (coliform organisms) 16-17, 20, 173
Coliphages 18
Colour 123, 180
Committed effective dose 115
Compliance, monitoring to ensure 109-110
Concrete pipes, corrosion 140-141
Consumers
 acceptability to *see* Acceptability
 taps, sampling from 110
Contingency plans 142
Copper 46, 174
 acceptability 125, 180
 corrosion 140
Corrosion 12, 138-142
 control strategies 142
 indices 141-142
 microbiological aspects 141
 pipe materials 140-141
 water composition effects 45, 139
*Crangonyx pseudogracilis* 12
*Cryptosporidium* 8, 15, 23
*Cryptosporidium parvum* 11
*Culex* larvae 12
Cyanide 46-47, 174
Cyanobacteria
 blooms 12
 toxins 9-11
Cyanogen chloride 105, 178
*Cyclops* 12
*Cylindrospermum* 11

2,4-D 81-82, 176
2,4-DB 91, 92, 176
DDT 80, 176
Dermal absorption, chemicals 31
Detergents, synthetic 129, 181
Dialkyltins 75, 175
1,2-Dibromo-3-chloropropane (DBCP) 81, 176
Dibromoacetonitrile 104, 178
Dibromochloromethane (DBCM) 99, 100, 177
1,2-Dibromoethane (ethylene dibromide) 83, 176
Dichloramine 94, 177

Dichloroacetic acid 102, 177
1,1-Dichloroacetone 103
Dichloroacetonitrile 104, 178
1,2-Dichlorobenzene 69, 129, 175, 181
1,3-Dichlorobenzene 69, 175
1,4-Dichlorobenzene 69-70, 129, 175, 181
Dichlorobenzenes (DCBs) 69-70, 129
1,1-Dichloroethane 59, 175
1,2-Dichloroethane 59, 175
1,1-Dichloroethene 61, 175
1,2-Dichloroethene 61-62, 175
Dichloromethane 58-59, 175
2,4-Dichlorophenol 97, 130, 177, 181
2,4-Dichlorophenoxyacetic acid (2,4-D) 81-82, 176
1,2-Dichloropropane 82, 176
1,3-Dichloropropane 82, 176
1,3-Dichloropropene 82-83, 176
Dichlorprop 91, 92, 176
Dieldrin 77, 176
Di(2-ethylhexyl)adipate (DEHA) 70-71, 175
Di(2-ethylhexyl)phthalate (DEHP) 71-72, 175
Disinfectant by-products 1, 7, 93-105
    acceptability 129-130, 181
    guideline values 177-178
    health risks 3-4
Disinfectants 93-105
    acceptability 129-130, 181
    guideline values 177-178
Disinfection 20, 133, 135-136, 137
    emergency 143
Distribution networks 137-138
    nuisance microbes 12, 137
Dose
    infective 10-11, 13
    radiation 115-121
*Dracunculus medinensis* 8, 12
*Dreissena polymorpha* 12

*Echinococcus* 9
Edetic acid (EDTA) 74, 175
Emergency measures 142-143
*Entamoeba histolytica* 8, 11
*Enterobacter* 16
*Enterobacter cloacae* 16-17
*Enterococcus* 17
Enteroviruses 10, 15, 23
Epichlorohydrin (ECH) 72-73, 175
*Escherichia* 16
*Escherichia coli*
    bacteriophages 18
    guideline values 173
    as indicator of faecal pollution 15, 19, 20, 22
    pathogenic 8, 10
    treatment effects 135

Ethenes, chlorinated 60-64, 175
Ethylbenzene 66, 128, 175, 180
Ethylene dibromide (EDB) 83, 176
Ethylenediaminetetraacetic acid (edetic acid) 74, 175

Faecal contamination 6, 8
    emergency measures 142-143
Faecal indicator organisms 14-19, 20
    methods of detection 18-19
Faecal streptococci 15, 17-18
*Fasciola* 9
*Fasciolopsis* 9
Fenoprop 91, 92, 176
Filtration
    point-of-use 135-136
    pre-treatment 133
    rapid 134-135
    slow sand 134-135
*Flavobacterium* 9
Flocculation 134
Fluoride 47, 174
    removal 136
Formaldehyde 98, 177

Galvanic cells 139
*Gammarus pulex* 12
Geosmin 12
*Giardia* 8, 15, 23
*Giardia intestinalis* 11
Ground water
    protection 132
    treatment 23, 24, 136
Guideline values
    chemicals 32-38, 174-179
    consumer acceptability 180-181
    microorganisms 13-14, 21-24, 173
    nature 4-6
    provisional 5-6, 178
    radionuclides 116-121, 179
    tables 172-181
Guinea worm (*Dracunculus medinensis*) 8, 12

Halogenated acetonitriles 103-105, 178
Hardness 48, 125, 180
Helminths 9, 11
Hepatitis A virus 10
Hepatitis E virus 10
Hepatitis viruses, non-A, non-B 10
Hepatolenticular degeneration 46
Heptachlor 83-84, 176
Heptachlor epoxide 83-84, 176
Hexachlorobenzene 84, 176
Hexachlorobutadiene (HCBD) 73-74, 175
$\gamma$-Hexachlorocyclohexane ($\gamma$-HCH; lindane) 85, 176

Hexachlorodibenzo-*p*-dioxin 88, 89
Hydrogen sulfide 48, 125-126, 180
Hypertension 55
Hypochlorite 135

Impoundments 133
Infective dose 10-11, 13
Infiltration, pre-treatment 133
Inhalation
   chemicals 31
   microorganisms 9
Inorganic chemicals 39-57
   acceptability 124-128, 180
   guideline values 174
Inspectorate, regulatory 28
International Agency for Research on Cancer (IARC) 30, 35, 36
International Commission on Radiological Protection (ICRP) 114, 119
International Organization for Standardization (ISO) standards 19, 26, 118-119
International Programme on Chemical Safety (IPCS) 30
Iodine 95-96, 177
Iodine-129 ($^{129}$I) 117, 118
Iodine-131 ($^{131}$I) 117, 118
Iron 48-49, 126, 180
Iron bacteria 126, 141
Isoproturon 84-85, 176

Joint FAO/WHO Expert Committee on Food Additives (JECFA) 30, 32-33
Joint FAO/WHO Meetings on Pesticide Residues (JMPR) 30, 32-33

*Klebsiella* 9, 16

Langelier index 140-141
Larson ratio 142
Lead 49-50, 174
   corrosion 140
   monitoring 108, 109-110
Lead-210 ($^{210}$Pb) 117, 118
*Legionella* spp. 9, 13
Lindane 85, 176
Lowest-observed-adverse-effect level (LOAEL) 32, 33, 34

Magnesium sulfate 127
Manganese 50-51, 126, 174, 180
MCPA 85-86, 176
MCPB 91, 92, 176
Mecoprop 91, 92, 176
Mercury 51, 174
Methaemoglobinaemia 53

Methoxychlor 86, 176
Methylene chloride (dichloromethane) 58-59, 175
2-Methylisoborneol 12
Methylmercury 51
Metolachlor 87, 176
Microorganisms
   corrosion due to 141
   criteria for selection 6
   in distribution networks 137, 138
   faecal indicators 14-19, 20
   guideline values 13-14, 21-24, 173
   health risks 3
   infectious 8-9, 10-11
   infective dose 10-11, 13
   nuisance 11-12, 137
   persistence in water 10-11, 12-13
   significant 8-14
   toxins 9-11
Microbiological aspects 8-29
Microbiological quality 93, 131-132
   guideline values 21-24, 173
   monitoring 24-29
   recommendations 20-24
   selection of treatment processes 20
   treatment objectives 21
*Microcystis* 11
*Microcystis aeruginosa* 11
Mineral waters, natural 6
Mixtures, chemical 39
Model, linearized multistage 38
Molinate 87, 176
Molybdenum 51-52, 174
Monitoring 131
   chemical constituents 105-113
      analysis 111-113
      sample collection 110-111
      sampling programme design 106-110
      to ensure compliance 109-110
   microbiological quality 24-29
      sampling frequencies 25-26
      sampling procedures 26-27
      surveillance programme requirements 28-29
Monochloramine 94, 177
Monochloroacetic acid 101-102, 177
Monochlorobenzene (MCB) 68-69, 129, 175, 181
Mosquito larvae 12
Multiple-barrier concept 14, 21, 133
MX (3-chloro-4-dichloromethyl-5-hydroxy-2(5H)-furanone) 98, 177
Mycobacteria, "slow-growing" 9

*Naegleria fowleri* 9, 13
*Nais* worms 12
National standards, developing 1, 2, 4, 5
*Necator* 9

Nematodes 12
Nephelometric turbidity unit (NTU) 124, 135
Nickel 52, 174
Nitrate 52-53, 174
Nitrilotriacetic acid (NTA) 74-75, 175
Nitrite 52-53, 174
No-observed-adverse-effect level (NOAEL) 32, 33, 34
*Nodularia* 11
Norwalk virus 10
*Nostoc* 11
Nuisance organisms 11-12, 137

Odour 122, 123, 180
Opportunistic pathogens 9
Organic chemicals 57-75
    acceptability 128-129, 180-181
    guideline values 175
Organotins 75
*Oscillatoria* 11
Oxygen, dissolved 53, 126, 139, 180
Ozonation 135, 136-137

Parasites 8, 9, 15
Parasitological quality guidelines 23-24
Passivation 139
Pendimethalin 87-88, 176
Pentachlorophenol (PCP) 88-89, 176
Permethrin 89, 176
Pesticides 75-93, 176
pH 53-54, 127, 180
    corrosion and 139, 142
    saturation 141
Physical parameters 123-124, 180
Pipes
    coal-tar linings 68
    corrosion 138-142
    quality of materials 7
*Plumatella* 12
Plumbosolvency 140
Plutonium-239 ($^{239}$Pu) 117, 118
Polonium-210 ($^{210}$Po) 117, 118
Polychlorinated dibenzo-*p*-dioxins (PCDDs) 88
Polychlorinated dibenzofurans (PCDFs) 88
Polynuclear aromatic hydrocarbons (PAHs) 67-68
Potassium-40 ($^{40}$K) 118
Pre-disinfection 133
Pre-treatment 133
Propanil 90, 176
Protection, water sources 2, 3, 132, 136
Protozoa 11
*Pseudomonas aeruginosa* 9, 10, 13
Pyridate 90, 176

Quality control, chemical analysis 111, 113

Radiation
    committed effective dose 115
    dose 115
    environmental exposure 114-115
    health effects 115
    reference level of dose
        practical application 116-121
        recommendations 115-116
Radionuclides 5, 114-121
    analytical methods 118-119
    guideline activity concentrations 116-121, 179
    health risks 4
    strategy for assessing drinking-water 119, 120
Radium-224 ($^{224}$Ra) 117, 118
Radium-226 ($^{226}$Ra) 116, 117, 118
Radium-228 ($^{228}$Ra) 117, 118
Radon ($^{222}$Rn) 117, 120-121
*Rahnella aquatilis* 17
Regulatory inspectorate 28
Reservoirs 133
Risk-benefit approach 2
Rotavirus 10
Rural areas, treatment processes 133, 136

*Salmonella* 8, 10
*Salmonella typhi* (typhoid fever) 10, 13
Samples
    analysis 111-113
    collection 110-111
    stability of chemicals 111
Sampling
    frequencies 25-26, 107
    locations 107-108, 109
    procedures 26-27
    programme for chemical constituents 106-110
    times 108-109
Schistosomiasis 9
Sedimentation 134
Selenium 54, 174
*Serratia* 9
*Serratia fonticola* 17
*Shigella* 8, 10
Silver 54-55, 179
    point-of-use filters 135-136
Simazine 90-91, 176
Skin cancer 41
Snails 12
Sodium 55, 127, 180
Sodium thiosulfate 27
Solids, total dissolved (TDS) 56, 127-128, 180
*Spirometra* 9
Stochastic effects 115
Streptococci, faecal 15, 17-18
*Streptococcus* 17-18
*Strongyloides* 9

Strontium-89 ($^{89}$Sr) 117, 118
Strontium-90 ($^{90}$Sr) 116, 117, 118
Styrene 66-67, 128-129, 175, 180
Sulfate 55-56, 127, 180
Sulfite-reducing clostridia 15, 18
Surface water
   protection 132
   treatment 23, 24, 136-137
Surveillance programmes 28-29

2,4,5-T 91, 92-93, 176
*Taenia solium* 9
Taps, sampling from 110
Taste 122, 123, 180
Temperature 123-124, 180
   microbial survival and 12-13
3,3',4,4'-Tetrachloroazobenzene (TCAB) 90
Tetrachloroethene 63-64, 175
Thermotolerant coliform bacteria *see*
   Coliform bacteria, thermotolerant
Thorium-232 ($^{232}$Th) 117, 118
Tin, inorganic 56, 179
Tolerable daily intake (TDI) 32-33
   allocation 34-35
   derivation of guideline values 32-35
Toluene 65, 128, 175, 180
Total dissolved solids (TDS) 56, 127-128, 180
Toxins, cyanobacterial 9-11
*Toxocara* 9
Treatment (*see also* Disinfection)
   chemical by-products 7
   choice of processes 20, 136-137
   objectives 21, 22
   processes 132-136
   virus reduction 23, 24
Tributyltin oxide (TBTO) 75, 175
Trichloramine 94
Trichloroacetaldehyde 103, 177
Trichloroacetic acid 1, 102-103, 177
Trichloroacetonitrile 103, 104, 105, 178
Trichlorobenzenes 70, 129, 175, 181
1,1,1-Trichloroethane 59-60, 175
Trichloroethene 62-63, 175
Trichloronitromethane 105
2,4,6-Trichlorophenol (2,4,6-TCP) 97-98, 130, 177, 181

*Trichuris* 9
Trifluralin 91, 176
Trihalomethanes 1, 99-101, 134, 177
Tritium ($^{3}$H) 117
True colour units (TCU) 123
Turbidity 124, 135, 180
   treatment objectives 21, 22
Typhoid fever (*Salmonella typhi*) 10, 13

Ultraviolet irradiation 135
Uncertainty factors 32, 33-34
Underground storage tanks 137
United Nations Scientific Committee on the
   Effects of Atomic Radiation (UNSCEAR) 114
Uranium 57, 174
Uranium-234 ($^{234}$U) 117, 118
Uranium-238 ($^{238}$U) 117, 118
Urban areas, treatment processes 132-133

*Vibrio cholerae* 8, 10
Vinyl chloride 60-61, 175
Vinylidene chloride (1,1-dichloroethene) 61, 175
Virological quality guidelines 23
Viruses 10
   small round 10
   treatment effects 135

Water
   consumption data 30-31
   hardness *see* Hardness
   intake, allocation 34-35
Water quality (*see also* Guideline values)
   monitoring *see* Monitoring
   protection and improvement 131-143
Water sources
   protection 2, 3, 132, 136
   selection 132
Waterborne infections 3, 8-9
   opportunistic/other water-associated
     pathogens 9
   orally transmitted 8, 10-11

Xylenes 65-66, 128, 175, 180

*Yersinia enterocolitica* 8, 10

Zinc 57, 74, 128, 180